Planning a Wilderness

Planning a Wilderness

Regenerating the Great Lakes Cutover Region

James Kates

Published in cooperation with the Center for American Places
Santa Fe, New Mexico, and Harrisonburg, Virginia

University of Minnesota Press
Minneapolis
London

The University of Minnesota Press gratefully acknowledges the generous assistance provided for the publication of this book by the Hamilton P. Traub University Press Fund.

Frontispiece. Visitors pose at the base of a white pine tree at Newald, Wisconsin, 1952. Photograph courtesy of State Historical Society of Wisconsin, WHi (X3) 26551.

Portions of this book were previously published in the following essays, and the University of Minnesota Press gratefully acknowledges permission to reprint the material here. "The Conservationist as Journalist: P. S. Lovejoy and the Fight for the Cutover," *American Journalism* 12, no. 2 (Spring 1995): 123–41; reprinted by permission of *American Journalism*. "James Oliver Curwood: Antimodernist in the Conservation Crusade," *The Michigan Historical Review* 24, no. 1 (Spring 1998): 73–102; reprinted by permission of *The Michigan Historical Review*.

Published by the University of Minnesota Press
111 Third Avenue South, Suite 290
Minneapolis, MN 55401-2520
http://www.upress.umn.edu

Library of Congress Cataloging-in-Publication Data

Kates, James.
    Planning a wilderness : regenerating the Great Lakes cutover region / James Kates.
        p.   cm.
    Includes bibliographical references and index.
    ISBN 0-8166-3579-X — ISBN 0-8166-3580-3 (pbk.)
    1. Reforestation—Great Lakes Region.   2. Forest management — Great Lakes Region.   I. Title.
    SD409.K28   2001
    634.9'56'0977 — dc21   00-011868

Printed in the United States of America on acid-free paper

The University of Minnesota is an equal-opportunity educator and employer.

11 10 09 08 07 06 05 04 03 02 01   10 9 8 7 6 5 4 3 2 1

I discovered another world. It had been there before, for long and long, but I had never seen nor felt it. All discoveries are made in that way: a man finds the new thing, not in nature but in himself.
— David Grayson (Ray Stannard Baker)
Adventures in Contentment, 1907

Contents

Acknowledgments

The writing of a book is a solitary effort, but the process of learning and reflection that underlies that writing is a communal journey. In the seven years of study, writing, and revision leading to this work, I was blessed with fine teachers, supportive colleagues, and a family without whose understanding and encouragement this book could not have taken shape.

My doctoral adviser at the University of Wisconsin–Madison, James L. Baughman, frequently likened his role to that of an attorney — telling me what I needed to know even when I might not want to hear it. His good counsel helped me to navigate the academic bureaucracy and, I hope, to be a better scholar. His professionalism and his sense of humor have always inspired me. He has a grateful client.

My committee members were similarly supportive. Stephen Vaughn's door is always open to students, literally and otherwise. His guidance helped me write a seminar paper on P. S. Lovejoy, which later became a conference paper, a journal article, and chapter 3 of this work. Summoned late in the game, Sharon Dunwoody graciously agreed to lend her expertise on science journalism. Paul Boyer, of the Department of History, helped shape my early thinking on the forestry question during his graduate seminar on intellectual history. And John Milton Cooper Jr., also of the Department of History, read an early draft, gave me some stylistic pointers, and offered his boundless enthusiasm for all things Progressive.

Other scholars have guided my thinking as well. Robert Gough of the University of Wisconsin–Eau Claire read the manuscript in its entirety, providing a provocative critique from a perspective quite different from my own. Charles Twining, a veteran of forest history, brought a similarly practiced eye to the job. Sidney Fine, of the University of Michigan, shepherded my first paper on James Oliver Curwood when I was in the master's program in history at Ann Arbor many years ago. His example of scholarly rigor is unmatched. William Cronon, an environmental historian at Madison, has been a constant inspiration. His melding of a fierce intelligence and a gentle demeanor is a formula I can only aspire to.

The UW–Madison School of Journalism and Mass Communication provided fellowship support, research and travel funding, and the opportunity to work as a teacher for several semesters. I am particularly grateful to Chairman-Professor Bob Drechsel and to graduate coordinator Susie Brandscheid, who has guided many students through the procedural thickets that line the path to the Ph.D.

My fellow graduate students offered constant support, not just intellectually but in terms of plain friendship. Special thanks are due to Katy Culver, Heather Hartwig Boyd, Inger Stole, Carolyn Bronstein, Sheila Webb, Andrew Feldman, and John Couper.

George F. Thompson and his staff at the Center for American Places took on this project while it was still a dissertation, helped me pare and focus the manuscript, and found a home for the book with the University of Minnesota Press. Without the Center's assistance I would still be wandering in the wilderness of academic publishing. At Minnesota, Mike Stoffel and Daniel Leary handled production duties; Anne Running did the copyediting.

This work reflects the assistance of librarians and archivists in three states and the District of Columbia. The Bentley Historical Library at the University of Michigan was my home for a month all told; two weeks there were underwritten by a Bordin-Gillette Researcher Travel Fellowship. The Bentley's Francis X. Blouin Jr., William Wallach, and Nancy Bartlett opened the library's riches to me and made me feel welcome. An Alfred D. Bell Travel Grant allowed me to spend a week in residence at the Forest History Society in Durham, North Carolina; I am grateful for the hospitality and assistance of the Society's Harold K. "Pete" Steen, Cheryl Oakes, and Ed McIntyre. The staff at the Library of Congress helped me make the most of a brief trip there, retrieving selected cartons of the voluminous Gifford Pinchot papers from remote storage in advance of my visit. Some of my ablest assistance came closest to home, at the University of Wisconsin campus libraries, at the UW–Madison Archives, and from Harry Miller and his staff at the Archives Division, State Historical Society of Wisconsin.

My first child, Margaret Paige Kates, was born just after the drafting of chapter 5. Having spent three weeks grappling with the ghost of James

Oliver Curwood, I desperately needed reminding that there are more important things in the world than the practice of history. Maggie has done that, and so much more. Lucy Rose Kates arrived during the long process of overhauling the dissertation into a book manuscript, again providing a sense of perspective and a shot of much-needed inspiration.

My mother, Elizabeth Kates, died as this book was going to press. I will treasure her memory.

My final and most heartfelt thanks are reserved for my wife, Sharon L. Schmeling. She has been a partner and a soul mate at each step in this long process. She has endured my arcane enthusiasms and my incessant grumblings. Suffice it to say that without her, the material, emotional, spiritual, and intellectual conditions under which this book took shape could not have existed. I dedicate every word of it to her, with love.

INTRODUCTION

The Writer and the "Lesser" Landscape

In the summer of 1924, fresh from the Republican convention in Cleveland, Theodore Roosevelt Jr. linked up with a Chicago acquaintance and rode a train north to Marinette County, Wisconsin, on the border with Upper Michigan. The two men fished for trout in rushing streams and for bass in the cool waters of Coleman Lake. The young Roosevelt, who had inherited his late father's gift for translating outdoor adventure into vigorous prose, recounted the trip zestfully for the genteel readers of *Scribner's Magazine.*[1]

To modern eyes, the most striking aspect of Roosevelt's tale is the tentative manner in which he describes the countryside. "Once it was covered with a great forest, but that has been lumbered off," he wrote. By the time of his visit, it was "very rare to find real woods in this country." Soils were poor, and the few farms established after the timber was gone had mostly failed. Ravaged by periodic wildfires, the North Country was hardly an untouched "wilderness"; instead, it was a land whose desolation bore witness to the fleeting presence and enduring folly of human beings. Like his father, Roosevelt professed that a man's temperament, and a nation's, could be shaped through an ongoing encounter with nature. The blackened acreage of northern Wisconsin was a far cry from the rugged vistas that had forged the American "spirit of sturdy self-reliance," he wrote. Still, Roosevelt sensed—however indistinctly—that some sort of collective action might transform this denuded landscape into a "reservoir of wealth and an invaluable playground for the people." With foresight, coordination, and planning, the "wilderness" might somehow be remade.[2]

Three-quarters of a century later, much of what Roosevelt dimly pro-phesied has come to pass. The former "cutover" of the northern Great Lakes region is growing trees again; its sparkling lakes draw a steady stream of visitors from the cities. The region's economy is still relatively anemic, relying on forest industries, tourism, and government, but it no longer is wracked by the boom-and-bust cycles that characterized the logging era and the agricultural experiment that followed. Roosevelt's vision of a "playground" has been realized more fully than he might have dreamed. To an extent that cannot be measured, the millions of per-son-hours spent fishing, hiking, and otherwise enjoying the landscape each year probably do contribute to a sort of collective revitalization and civic health.

But as Roosevelt realized, none of this was inevitable. The new forest would be an engineered landscape; it would exist because its existence was, at some point, deemed desirable by persons in power. Though a product of nature, it would be not so much "natural" as naturalistic. The experts who built it would take cues from nature, but they fully in-tended that the finished product should be better than nature itself—more productive, more stable economically and socially, and more predictable in terms of its outputs. It was a bold experiment, one of numerous efforts at cooperative social engineering that were afoot dur-ing the 1920s. Unlike the factory floor or the planned cityscape, however, the forest must have seemed an odd place for such a blunt assertion of human will. The methods for growing it, the laws that encouraged its planting, the economic and political rationale for its protection—not only would all of these have to be "invented," but they would have to be explained and justified to the public as well. The naturalistic forest would be built by technical specialists: foresters, land economists, game man-agers, regional planners. At its base, though, the forestry question would be phrased in the broadest cultural terms: What was so important about trees that they had to be construed as indispensable to the American way of life?

Much of this work would be done by writers. Many of these writers were connected with the scientific or policy apparatus in state and federal governments, giving them a technical base with which to make their ar-guments. Most of them espoused the New Era's faith in planning, believ-ing that centralized intelligence could be brought to bear on the work-ings of nature to yield a steady harvest of economic and social benefits.[3]

On its face, this work was highly technical, involving the projection of societal goals onto a formerly uncoordinated landscape through the forces of expertise acting in three main areas: law, science, and administration. But a few gifted planners went beyond the dry certitude of science, into the realm of imaginative projection of the sort described by Coleridge.[4] The people who reimagined the cutover would not only have to dream of the future forest, but would also have to assemble an interlocking set of technical, political, legal, and economic inventions to bring it about. The best of them — Wisconsin's Aldo Leopold comes immediately to mind — possessed an uncanny ability to operate simultaneously in the ethereal and practical realms of conservation work. The essence of the dream lay in the details.

By 1933, the building blocks of reforestation were in place. Ranging from fire protection to rural zoning, these innovations were designed to make the cutover continuously productive. The desired outputs included not just wood, but also less tangible products such as recreation and social harmony. This intensive engineering effort would shape and channel not only the biological processes of the forest, but also the activities of the people who worked and played within it. For timber cutters, for hunters and anglers, and for those hoping to establish homes, the encounter with nature would no longer be a matter of individual whim, but would be overseen, facilitated, and sometimes restricted by the state.

On its face, the remade landscape would retain certain elements of the bygone frontier. Amid its groomed forests, citizens might discover a mythical touchstone to the pioneer past, a quality Leopold and others considered essential to the national psyche. But in the regrown cutover, the interaction of human beings and nature would be circumscribed to prevent exactly the sort of wasteful free-for-all that had marked the pioneer era. The new frontier might be inviting, but it was essentially synthetic.

Parish Storrs Lovejoy, a forester and magazine writer who was one of the first educated observers to look skeptically at the cutover situation, toured widely in the region during the summer of 1919. The Great Lakes cutover was huge — comprising more than fifty thousand square miles in northern Michigan, Wisconsin, and Minnesota — but Lovejoy soon discovered a distressing sameness about the landscape. Each time he topped a hill, he invariably would find himself looking across a sickening expanse of devastated acreage. Lovejoy, whose stated mission was to bring order and economic rationality to questions of land utilization, was unnerved by the haphazard nature of pioneer homesteading. "In all the country before and behind us there was nothing to see but the black snags and stumps and the popple brush," he reported at one stop, "save, far off, a few scattered and irregular spots which marked the clearings where the settlers were conducting their pursuit of life, liberty and happiness."[5] Rather than blossoming with farms, the vast majority of the cutover lands were "idle and desolate, an economic liability and a menace to life, property and prosperity because of the ever-recurring fires which sweep them bare, year after year."[6] Convinced that individualism had led only to ruin, Lovejoy was not the only observer to liken the cutover to the battlefields of Europe.

Even when remade, the cutover would lack the monumental scale and mythic resonance of the American West. The white pines, the centuries-old titans whose ranks had dominated the former Great Lakes forest, had been all but wiped out by loggers; in their place would rise smaller species that would be cut at intervals of a few dozen years. In areas where trees had been deliberately replanted, the neat rows were just one more reminder that the cutover landscape — destroyed and then regrown at the hands of human beings — would always be somehow "artificial."

Perhaps it is little wonder, then, that the Great Lakes forest has drawn comparatively scant attention over time from nature writers, national-level activists, and historians. John Muir, who reveled in the outdoors of central Wisconsin during his boyhood, nonetheless found the landscape of his dreams — and his livelihood — two thousand miles westward, at Yosemite. Gifford Pinchot, an eastern patrician whose ethos of "wise use" put him at odds with Muir in the early twentieth century, likewise refined his ideas against a backdrop of wide-open western spaces. Pinchot's shadow still falls heavily on the practice of forest history, as it does on forestry itself. While studies of the Great Lakes forest are not unknown, they are relatively few,[7] and they are dwarfed by a huge literature relating to the fierce conflicts — such as preservation versus use — waged by scientists and politicians over the forests of the West.

This is regrettable. The saga of the Great Lakes forest is (or should be) of interest to historians specifically because the region's woodlands had to be physically remade, rather than merely preserved or managed.

The story of the West, which predates the Great Lakes forest movement by a generation, entails many battles over existing resources, such as minerals, water, and old-growth timber. By contrast, the Great Lakes forest was logged in earnest beginning about 1865 and was largely depleted by 1910. It had to be regrown even as a myriad of interests—particularly agriculture—insisted that the land be put to other uses. As will be seen, many of the protagonists of this book spent their formative days in the West, and they incorporated western ideals and imagery into their campaign to remake the midwestern woods. But because they were working, literally, from the ground upward, they had to craft an economic, social, political, and spiritual rationale for the forest even as they went about the task of physical cultivation.

Similar work had attended the forest campaigns of the West, of course. But the Great Lakes crusade would be played out largely in the 1920s, a decade marked by the emergence of public relations and modern advertising—which sought to sell physical goods through association with less tangible dreams and desires. Forestry activists, such as those at the American Forestry Association who dreamed up countless ways to link trees with solitude, solace, and even patriotism and good citizenship, were not being entirely cynical when they suggested that forestry in the 1920s would best be sold to the public through "psychological" appeals. At its core, conservation was a conservative revolution, seeking to harness nature to provide a predictable stream of raw materials for industry and the type of land use that was thought conducive to social order, prosperity, and stability. Like the larger management ethos of the 1920s, it abhorred waste, encouraged cooperation, sought a middle way between state collectivism and unrestrained individualism, and placed its ultimate faith in a supposedly apolitical science of social and economic engineering. But the more prescient of the 1920s "experts" also recognized that conservation had to be pitched to the individual citizen as never before. Thus the centralized, knowledge-driven, bureaucratic work of 1920s conservation—such as reforestation—would be packaged and sold to the public in terms of individual wish fulfillment. These aspirations included everything from the psychic benefit of a walk in the woods to the yearning for a recently vanished American frontier. Conservation during the 1920s was still focused on efficiencies of production. But the rationale that was used to promote conservation—disingenuous though it may have been—contained the seeds of the modern environmental ethic focusing on consumption, social amenities (at least for those who could afford them), and quality of life.[8]

With a more or less chronological pattern, this book traces the increasing synergies between technical forestry, social engineering, and conscious mythmaking that marked Great Lakes reforestation. Chapter 1 describes the imperative for continuous production embraced by the forestry profession as part of the larger ascendance of planning after World War I. Chapter 2 outlines the crisis of American agriculture during the 1920s and the battle to rationalize patterns of land use, a drive that eventually would place much of the cutover off-limits to farming. Chapter 3, on the work of P. S. Lovejoy, spells out the economic prescription for the cutover, a plan that was far-reaching yet ultimately limited by its refusal to recognize the mythical value of the forest. Chapters 4 and 5 outline that mythical aspect, as seen in the glorification of the forest ranger and the spinning of forest fables in popular fiction.

From there, the remainder of the book tells of the sometimes clumsy melding of mechanism and myth that would propel Great Lakes reforestation into the New Deal years. As described in chapter 6, the remade landscape offered recreationists a facsimile of frontier experience, through the agency of state and private management backed by science. Chapter 7 profiles Harold Titus, a writer and conservationist who early on recognized the serviceability of myth in promoting rational development of the cutover region. Fire, whose presence in the woods was subject to a vast array of social and scientific interpretations, is the subject of chapter 8. Rural zoning, which barred human settlement on millions of cutover acres, was the capstone of cutover planning. Chapter 9 discusses the genesis and the impact of this far-reaching scheme, which ratified the cutover revolution by creating a synthetic frontier in the once-and-future forest.

There is another argument for a Great Lakes forest history. The powerful imagery of the West is closely tied to the modern-day wilderness movement, which promotes the preservation of large, unbroken tracts of land and water in something approximating their primeval condition. William Cronon has written that the wilderness ethic, as potent and pervasive as it has been in the larger environmental movement, has tended to further the dangerous, and illusory, idea of a stark dualism between man and nature. If human beings are viewed as intruders in the "natural" landscape, it is easy enough to view any human-touched landscape as beyond redemption. Government-designated "wilderness" areas are a spectacular national asset, with the Great Lakes region boasting one of the finest, in the far reaches of Minnesota on the border with Ontario. Still, a focus on preserving remote landscapes — while not without its ideological merits and laudable results — can in fact detract much-needed attention from the building of landscapes closer to home, where most of us spend most of our days.[9] Attractive but hardly pristine, the majority of today's Great Lakes forest is a "wilderness" on a human scale: a somewhat wild-looking landscape where every acre is a product of human intent.

Of course, the trampling feet of humanity bring all sorts of perils with them — of misplaced scientific judgment, environmental hubris, and a certain mean-spiritedness masquerading as a drive for social betterment. All of these shortcomings were evident in the Great Lakes forest movement during the 1920s; muted outcroppings of them can be seen worldwide to this day.[10]

But human foibles should not constitute a reason to deny the human place in nature. Indeed, in any human community, the social construction of nature determines the physical shape of the landscape, and the uses — tangible and otherwise — to which the land will be put. In the United States, whose immense natural resources are equaled only by its capacity to consume and destroy them, the preservation, management, and replanting of forests have always depended on shifting human conceptions of what the forest is good for. Like it or not, essayist Michael Pollan has written, "our metaphors about trees by and large determine the fate of trees."[11] The purpose of this book is to argue, as Pollan does, that such metaphors matter — indeed, that they are our most powerful means of maintaining a livable environment.

CHAPTER ONE

"Timber Famine," the Quest for Production, and the Lingering Frontier

> It has always seemed to me that the turn in our forest economy from the timber *mine* to the timber *crop* came in the decade ending with 1930. . . . The old migrant, "King Timber," and his slogan, "The cheapest source of raw material," were challenged on their own ground. The little trees were taking over.
> —WILLIAM B. GREELEY, *Forests and Men,* 1951

Increase Lapham saw it coming. A versatile and venerable Wisconsin scientist, he traversed that state's woodlands in the days after the Civil War and issued a dire warning: if the cutting of timber continued apace, much of Wisconsin would become a windblown desert and its people would be reduced to poverty. Lapham's *Report on the Disastrous Effects of the Destruction of Forest Trees Now Going On So Rapidly in the State of Wisconsin,* issued in 1867, was a skillful jeremiad, melding economic arguments with the nascent environmentalism of writers such as George Perkins Marsh. Forests, Lapham argued, were the bedrock of American civilization; not only did they supply raw materials to industry, but they nurtured agriculture by curbing erosion and flooding. If Wisconsin was to prosper, he wrote, it would have to subdue and manage its woods, cultivating them for sustainable yield. Otherwise, "the winds and droughts shall reduce the plains of Wisconsin to the condition of Asia Minor. Trees alone can save us from such a fate."[1]

Lapham's words were drowned out by sawmills. The wholesale clearing of hardwoods from southern Wisconsin farms, which had aroused his attention as early as 1854, suddenly was being dwarfed by a logging

1

boom in the Great Lakes pinelands. These sandy acres — encompassing the northern third of Wisconsin, the northeastern third of Minnesota, the entire Michigan Upper Peninsula, and nearly half of the Lower Peninsula — contained a staggering biological treasure, literally a gift of centuries. Interlaced with wetlands and brush were massive stands of virgin white pine — trees that grew to 150 feet or more, so fat at the trunk that it took three men to reach around one. Michigan doubled its lumber production from 1870 to 1880, eclipsing Pennsylvania, the former leader. One hundred and twelve steam sawmills lined the Saginaw River between Saginaw and Bay City, a stretch of just a dozen miles; in 1882 these mills pumped out a billion board feet of lumber and had so much timber of lesser quality on hand that they managed to fabricate three hundred million wooden shingles, as well.[2] In 1890 the Great Lakes forest supplied more than a third of the nation's lumber, its vast riches ferried westward on rail cars to build houses, barns, churches, and schools on the treeless Great Plains.[3] As Michigan's forests waned, Wisconsin took the lead, peaking at 3.4 billion board feet in 1899.[4]

Then the bottom fell out. Great Lakes loggers would continue cutting the forest, but the quantity and quality of their output slumped precipitously. With the choicest timber gone, crews were relegated to cutting scattered hardwood stands, or a few scruffy species contemptuously known as "weed trees." Rivers that once had been choked with logs ran free again, and one by one, the great sawmills — at Saginaw, Michigan; Chippewa Falls, Wisconsin; Stillwater, Minnesota; and elsewhere — fell silent. By 1929 the region's annual cut would fall by more than 70 percent. In one commentator's assessment, the noble white pine, which had been the symbol and staple of an entire woodland economy, was now a "vanishing American."[5] But the real sea change had come some years earlier, about the time of World War I. That's when the denuded pinelands, once commonly referred to as an "empire," acquired a more timely and less grandiose sobriquet: the "cutover."

On April 16, 1919, the chief forester of the United States addressed an uneasy crowd of lumbermen at their annual convention in Chicago. As the timber titans shifted in their seats, Henry S. Graves declared that the cut-and-run practices of the frontier past were obsolete. The problems associated with forest depletion — timber shortages, hardships to industry and consumers, and the flight of capital from cutover forest

regions—constituted an emergency that could "gravely affect the national welfare." Americans, Graves declared, had to begin growing trees
as a crop. But because a timber crop took half a century to mature, the
present crisis demanded foresight and centralized coordination on an
unprecedented scale. The massive government giveaway of timberlands
in the nineteenth century had been a "mistake," fueling waste and spec-
ulation even as it opened the continent to settlement. Now, in the name
of patriotism, the timber barons had to atone for their sins by tending
the forests of America, three-quarters of which remained in private
hands. Graves spoke of "cooperation," but behind this evasive term lay
the specter of government compulsion—even confiscation.[6]

A few months later, a special committee of the Society of American
Foresters endorsed Graves's sentiments and made plain his veiled threats.
At the helm of the panel was Gifford Pinchot, the patrician crusader
who had defined the moral and organizational character of early forestry.
As chief forester under Theodore Roosevelt, Pinchot had fought a years-
long battle to establish federal primacy on federal lands, only to be pil-
loried and politically bruised under the more cautious administration
of William Howard Taft. An "incorrigible" activist, in the estimation of
one historian, Pinchot had overreached his authority and helped to bring
about the "spectacular collapse" of his Forest Service career.[7] Now, as
footloose critic and father figure to a generation of foresters, he could
speak far more freely than Graves dared. The citizens of the United States,
Pinchot declared in the committee's report, were consuming nearly three
times as much timber as they grew. Private ownership of timber lands
carried with it a unique public trust, one that could be enforced by government.[8] Lumbermen, Pinchot stated during an ensuing publicity blitz,
seemed immune to the powers of persuasion: "Uniform, nation-wide,
compulsory legislation is the only adequate remedy."[9] A new age of organization was dawning, and the mossbacks of the lumber business had
to recognize it—both for the public interest and their own long-term
good. Lumbermen would have to be forced to do what other, more
enlightened business people were doing voluntarily: "The degree of
public regulation proposed is decidedly mild when compared with the
tremendous changes in the structure of industry now gradually assuming definite form throughout the whole world."[10] If lumbermen would
not enter the twentieth century voluntarily, Pinchot implied, the American people would have to drag them into it.

An atmosphere of moral crusade attached to the taking of trees was nothing new, of course. Early in the century, Pinchot and Roosevelt had locked up more than 100 million acres of standing timber, mainly in western forest reserves. On a single day in March 1907, for example, TR had created twenty-one new national forests in six states.[11] The rhetoric undergirding these efforts was an amalgam of Progressive ideas: efficiency, stewardship, and, especially, the need for a countervailing power of government to meet the threats posed by monopoly capital. In an age before mass tourism, forests were an abstraction in the lives of many Americans; thus TR and Pinchot sought to imbue trees with human values that could be appreciated by urbanites and flatland farmers. The Progressive Era forest was touted as a timber supply and as a site for dams or recreation, but it also was a measuring stick for the appealing yet vague tenets of the "Square Deal." What the forest *stood for* was at least as important as what one could *do* with it. Speaking to the American Forest Congress in 1905, Roosevelt had lambasted the lumbermen who "skin the country and go somewhere else . . . whose idea of developing the country is to cut every stick of timber off of it, and leave a barren desert." With an equal mix of policy and pure bluster, TR's forestry program was just one of many that used government power to manage the widening gap between individual ambition and the greater good.[12]

The forest crusade circa 1919 was different. Under Roosevelt, Pinchot had asserted a government interest in standing timber and had developed the rudiments of selective harvest and law enforcement by a newly professionalized ranger corps. The challenge now was to grow new trees where old ones had been cut. The new forestry was dauntingly complicated. Who would own the timberlands? Who would control them? What rationale would be used to "sell" forestry to a skeptical public in a conservative era? And who would develop the scientific knowledge and accounting procedures for growing a crop that took a lifetime to mature?

Pinchot had repeatedly invoked the specter of a "timber famine" during his days with Roosevelt; now he revived it to energize a new forestry movement that was not just political or moral but organizational. The forestry of the 1920s would represent a complex value calculus, in which information and expertise would be brought to bear on public and private lands in the name of desired social ends. Besides mere board feet of timber, acres could be actively managed to yield social stability, economic predictability, and long-term prosperity. Achieving those goals would require a complex nexus of persuasion, government assistance, economic incentive, and (perhaps) compulsion. A few firebrands, such as Pinchot, might brazenly wave the big stick, but the dominant model during the 1920s was one of cooperation. Herbert Hoover, who as commerce secretary became the decade's premier apostle of planning, feared that the unrestrained, uncoordinated taking of natural resources would not just hobble industry, but would undermine the recreational opportunities that were vital to the American spirit. Government, in Hoover's view, would become the locus of a new, expert stewardship — doing research, disseminating data, and generally goading producers toward enlightened self-interest. Underlying this philosophy was an idea that had already been tested and proved in Europe: that nature itself could be groomed, managed and made continuously productive.[13]

Defining the "Famine"

Was the United States, circa 1920, really running out of wood? And if it was, could ordinary citizens be made to care? How could forestry, seemingly a technical pursuit of interest primarily to industry, be translated into a matter of broad national urgency?

The answer, as Pinchot and a few others realized, was to frame the forestry question in terms far more expansive than board feet. Trees themselves were tangible objects, but a forest was an abstraction, something so big that it had to be conceived of metaphorically. The Pinchot-Roosevelt forest of Progressive days had not been preserved for its own sake — indeed, Pinchot believed that the only real measure of conservation was its utility to human beings. The woods could be cut, but cut only gradually, at a rate calculated to supply timber in perpetuity. If the timber titans wouldn't make this happen, the federal government would, by nationalizing trees and rationing their harvest. Simply stated, the forests of the West had been a metaphor for Progressive beliefs about the larger economy. As chief forester, Pinchot had been less a scientist than a broker of ideas. His work had been a high-stakes game of power and persuasion, supported by an aggressive publicity apparatus that presented forest questions in metaphorical terms.[14]

By 1920, Pinchot was well past his prime as a forester; his conception of the woods as an arena for the flexing of Progressive muscle had been supplanted by narrower technical endeavors, such as land economics, that he understood only vaguely. Still, many of the men who had been greenhorns in Pinchot's Forest Service had risen to positions of influence in conservation—as state and federal foresters, professors, or writers. The tireless "G.P." continued as mentor to these men, issuing a constant stream of encouragement and advice. He still recognized, as keenly as he had in his days with TR, that publicity would be the prime mover of the remade forest.

As if to prove that his skills were still sharp, Pinchot helped to coin one of the most pervasive forest metaphors of the 1920s: the image of the managed woodland as a repository of permanence. In Progressive days, "permanence" generally had implied a reliable supply of timber for industry, but the concept would prove readily expandable. The products of the groomed forest, as Pinchot now conceived of them, would include not just timber, but steady employment, ready capital, and the intangible brand of hope—even outright happiness—that could come only from living in a stable community with an assured future. Pinchot was hardly the first to liken the new forest to a garden. The regrown forest would be a thoroughly domesticated landscape, one that harnessed the workings of nature for explicitly human ends: "The forests which will be raised from now on will not be tangles of wilderness, left alone for a century or so and then ripped off so as to leave the country desolate and poor. Instead, they will be carefully tended and protected and, once established, will be permanently productive. Work in the forests will become a regular and permanent business."[15]

Other writers amplified the theme. Some wrote for general-interest magazines and newspapers, others for forestry periodicals, striving to forge a new organizational ethos for a fledgling profession. Many invoked images of the world war to describe how the cutover landscape had been stripped bare, ravaged by fire, and transformed into a desolate moonscape. Reaching backward for a pioneer ideal, more than one writer declared that government expertise would help citizens to "make homes" in forest regions. The forest landscape, under this scheme, could be managed to yield community stability as a by-product of economic growth. (As will be seen, the idea of "making homes" was especially malleable, for it also could be used *against* foresters on the assumption that the propagation of trees prevented potential settlers from securing sufficient land.) Russell Watson, a young forestry professor at the University of Michigan, viewed waste as a moral evil and unchecked logging as nothing less than the "devastation" of an entire region and its inhabitants. Only through the supervision of experts, Watson believed, could the cutover be made to yield long-term opportunity for the individual: "The primary duty of government is to administer its territory in such a way as to make it a permanent home of comfort and well-being for its people."[16] Watson was writing in the *Journal of Forestry;* his article was one of many in which foresters struggled to articulate a social rationale for their work in the 1920s. But how could this scheme be advanced to a wider audience?

One of the more remarkable champions of the new forestry was an articulate and well-born Hoosier named Ovid M. Butler. Like many of his colleagues, Butler was an alumnus of Pinchot's tight-knit conservation corps, having served at forest outposts in Idaho and Utah.[17] About 1918 he went to work at the U.S. Department of Agriculture's Forest Products Laboratory in Madison, Wisconsin, which was working to find uses for forest waste products. There, Butler was infused with a gospel of efficiency in the utilization of forest crops. A vivid writer, he soon was extolling the wonders of such products as turpentine and paper bags. Science, he told a visiting group of southern pine salesmen, was an "Aladdin's lamp" that could conjure value from forest products previously thought to be worthless.[18] In Butler's inventive mind, this seemingly narrow enthusiasm soon burst its bounds. In 1923 he became resident editor for the American Forestry Association, an industry-led group that advocated the management of forests for sustained yield. Butler quickly proceeded to remake the association's magazine, *American Forestry,* as a publication that celebrated the forest's connection to the American way of life.[19]

At first, the connection as Butler saw it was mostly material. In an early article, he set up a running fictional dialogue between an economist and a traveling salesman who were riding through a cutover forest in a train. The drummer, an obnoxious sort, mourned the forest not one bit. But the economist won his case by noting the presence of wood products in all sorts of places—not just in housing, but in everyday products ranging from phonograph records to soap to linoleum. The prosperity of the United States, Butler's economist declared, was based on an almost insatiable "forest hunger," and the nation would ignore that hunger at its peril. To feed it, Americans had to plant crops of trees just as they did fields of wheat.[20] Five months later, Pinchot himself weighed in, bemoaning the "tragedy" that had overtaken the forest hamlets of Pennsylvania, where he had served as chief forester in 1920–22 and now was governor. The little boom town of Norwich, for example, seemed destined to become a ghost town, as a sawmill had come and gone within ten years: "Discouragement and despair are written everywhere in the village."[21]

Under the rubric of continuous production, forestry was a handmaiden to industry — indeed, an industry in itself. As opposed to timber "miners," who logged mature trees and then moved on, industrial foresters sought stable and predictable yields from long-term land holdings. They were motivated, not by altruism, but by a simple truth: available virgin timber, which in economic terms was essentially a free good, was becoming scarce. Butler recognized that industrial foresters and the forestry profession were not antagonists but natural allies. Only a managed, continuously cultivated forest would require the sort of experts being churned out by the nation's forestry schools. The most visionary businessmen, in Butler's eyes, were beginning to regard trees as a form of long-term capital investment, which would return dividends in perpetuity if properly tended. His allies in this quest extended all the way to Herbert Hoover, who as commerce secretary was seeking voluntaristic, though government-guided, measures by which timbermen might substitute true "forestry" for wasteful "forest wrecking."[22]

In 1922, Butler had journeyed to the western Upper Peninsula of Michigan, where he viewed the vast hardwood forest recently purchased by the Ford Motor Company. Encompassing several hundred thousand acres, Ford's forest was run like a factory. Sixty lumberjacks — earning the famous Ford wage of five dollars a day — lived in orderly, electrically illuminated bunkhouses. A modern sawmill utilized mass-production techniques, including a conveyor to move logs in and lumber out. Henry Ford, a spokesman made clear, had no sentimental interest in trees. Rather, he was concerned about running out of timber for car production. The making of a single vehicle required 250 board feet of lumber. With sound management and selective cutting, the Ford forest could last forever. The company was managing trees the same way it managed people — with a thorough bureaucratic discipline, a limited measure of growing space, and paternalistic oversight to ensure permanent prosperity. Butler, clearly impressed with this sylvan assembly line, noted only one sign of dissension: the Ford lumberjacks were not allowed to spit on the floor in the clubhouse, a prohibition so odious to some that they quit their jobs.[23]

Still, it was easier to stir admiration for one corporate forester than it was to awaken the public to a general crisis in the woods. To complicate matters, the question as to whether America would "run out" of wood was probably misplaced. Annual per capita lumber consumption fell from 539 board feet in 1900 to 325 board feet in 1920 and continued to decline afterward.[24] A run-up in lumber prices after 1919 helped accelerate the search for alternatives to old-growth timber. The fruits of this search, many of them developed at the Forest Products Laboratory, would include plywood, pressboard, and advances in wood chemistry.[25] A revolution in plastics and synthetic fibers was in the offing. By 1930, Wisconsin forester-writer Aldo Leopold would recognize that the "famine" was more a question of "quality than quantity," with new manufactured wood products competing against wood substitutes for a role in industry. Meanwhile, there could just as easily be a "famine" in recreational areas.[26] The question, then, was not so much one of "running out" of wood. Rather, as Pinchot had perceived in his essay on ghost towns, it was a matter of the quality of American life and the part that forests played in the modern American existence.

A Change in Tactics

Just two years into his new life as an editor, Ovid Butler understood that fact with particular clarity. In 1925 he stood before an audience of foresters and confessed that, like them, he had expected the public to "seize upon our profession as a sort of savior from economic disaster." But the ploy had largely failed: the specter of a "timber famine" was simply too abstract and remote for the average American to care about. "So long as we visualize forestry in the public mind as a sort of machine for the production of wood in the distant forests, we are blind to our opportunities." Citing the public-relations guru Edward Bernays, Butler opined that foresters' opportunities were limited not so much by technical barriers as by the public perception of forestry. Foresters ultimately would do what the public *allowed* them to do, by way of political consent and government funding. The forest had to be "translated in popular and human terms" if foresters were to win support for their work. The forest experience had to be made immediate and real to the average person. If the dream of continuous production had failed to stir the public mind, then forest advocates would have to tout a more intimate brand of woodland permanence: "animal life, bird life, hunting, fishing, hiking, stream purity." These less tangible forest products, formerly considered "incidental" to the business of growing trees, had become "the open sesame to the public mind."[27] In the hands of a skilled publicist, the means used to "sell" the forest were almost infinitely malleable. Pinchot, for example, helped put an alluring spin on one writer's story by noting that revenues from state forests in Pennsylvania were earmarked for education. Thus the dry details of woodland economics and reforestation took on new life under the title "Planting Trees to Serve Our Children."[28]

By the mid-1920s, Butler and a number of other writers were working deliberately to reshape the American forest as a mythic space—a place celebrated not just for its documented history or its utility, but for the myriad associations that it might conjure in the public mind. Tellingly, just after going to work for the American Forestry Association, Butler had changed its magazine's title from *American Forestry* to *American Forests and Forest Life.* The remade publication took a decidedly "softer" approach, with features on hiking and camping, children's stories and poems, even pieces invoking spirituality and American Indian legends. The drumbeat for continuous production had not disappeared; it had simply been cloaked in themes that might prove more compelling to the periodical's broad readership. Butler's forest was a metaphorical mother lode. It was a link to the Old West and a repository of clean masculine virtue. It was a sanctuary for joyful recreation, contemplation, and self-discovery. In a territory still mostly uncharted, especially by professional foresters, it also could be a distinctly feminized space, not unlike a flower garden, where children might study nature's marvels. Butler's melding of Bernays with board feet certainly was more self-conscious than that of his fellow forester-writers, but it was quite in keeping with Madison Avenue. Much like an ad man selling automobiles or cigarettes, he "sold" the forest by linking the tangible product with less tangible fantasies, dreams, and desires. The underlying goal, of course, was to tame and rationalize every acre by bringing it under the regimen of scientific forestry. Such work inevitably would mean centralization of authority and the imposition of a management ethos in a place formerly conceived of as highly individualistic. But Butler and his colleagues were shrewd salesmen, assuring readers that the managed forest would provide Americans with endless opportunity for self-fulfillment— what one historian has called a "buffer against the effects of modern impersonalities of scale."[29]

One might even say that the forest itself had to be "invented." Its very existence in the American mind rested mostly on ideas of human utility, but circa 1922, those ideas were very much in flux. Frank A. Waugh, a prominent writer on landscape architecture, believed that foresters had made a serious mistake by touting only the direct economic benefits of their work. If a forest was defined by its human uses, certainly there was room for "fishing and camping and scenery" in the definition. In tune with an emerging consumer society, Waugh even speculated that those commodities might prove to be more valuable than timber. There was a certain affinity, he noted, between landscape architects and foresters. Landscape architects were accustomed to fashioning environments from scratch to serve a range of human needs, including privacy, relaxation, and even historical association and nostalgic yearning. Waugh recognized — more acutely than any technical forester could — that forestry advocates had to create a uniquely American idea of the forest, one that included links to the frontier and democratic participation in forest recreation.[30]

As Butler and Waugh both would come to realize, the American forest idea would evolve on two parallel tracks during the 1920s. One involved expanding notions of utility and specific forest uses, from lumbering to recreation and nature study. The other was the "mythical," specifically the popular historical and democratic ideas that would come to be associated with the forest experience. Myth and fact were sometimes at odds, particularly when the remade forest — a product of concentrated social and economic engineering — was celebrated as the last refuge for the individualistic wanderer. But much like the lore of the Old West, forest mythology would assume an importance above mere fact, in time becoming what Henry Nash Smith has termed a "historical truth" of its own.[31] The fact that the new forest was not a frontier, not a wilderness, not a proving ground for the pioneer, does not diminish the fact that many people conceived of it precisely in those terms.

Despite its allure, this mythic formula never translated into drastic political action. For one thing, compulsion on private forest land would prove neither politically palatable nor entirely necessary. Forestry work in the 1920s would be characterized by cooperative efforts, such as the fire-fighting structure set up by the Clarke-McNary Act of 1924. An agricultural depression after 1920 and a consequent crisis in land use would ensure that millions of acres reverted to forest, either in govern-

ment or private hands. Instead of timber supply or consumer conven-
ience, the rationale for forestry in the later 1920s was just as likely to be
camping or fishing or the spiritual refreshment to be found during a
walk in the woods. As Ovid Butler had discovered, the new forestry could
not be "sold" to the public on material terms alone. But the dream of
continuous production would hardly disappear from the woods. Rather,
it would be employed to yield an entirely new crop of social benefits as
the 1920s progressed.

Useful as it was to forest advocates, the image of the sturdy pioneer would
prove double-edged. Anyone who looked beneath the veneer of woodland myth would discover that the remade forest relied heavily on organization, expertise, and outside capital, all of which were inimical to
the lore of pioneer individualism. Henry Ford notwithstanding, the ideal
of continuous production might seem ill-suited to the forest — an environment that supposedly epitomized the raw struggle between man
and wilderness.

For advocates of forestry and other facets of a planned society, the
frontier myth was an ongoing challenge. Pinchot, like most foresters,
had been accused repeatedly of thwarting the pioneer spirit. In a particularly nasty Senate hearing in 1907, an Oregon lawmaker had lambasted
the "dreamers and theorists" of the Forest Service. The hare-brained
schemes of Washington eggheads were depriving the "lowly pioneer" of
his right to erect a "humble cabin" in "the shadow of the whispering
pines," the senator fumed.[32] Years later, through a mist of flawed hind-
sight, Pinchot recalled that most of his opposition had come from "grab-
bers," the big interests who were intent on raping the public domain
before ordinary settlers could move into it. The little man, he said, had
come around as soon as he realized that the Forest Service was on his
side: "Many settlers fought us in the beginning who afterward became
our steadfast friends."[33] But resentment of foresters was not confined
to silver-spurred ranchers or robber barons in top hats. Parish Storrs
Lovejoy, a young recruit in Pinchot's forest army, recalled an unusually
blunt display of public opinion one evening in 1906. In a bunkhouse
brimming with drunken lumberjacks, one rough-hewn character had
excoriated foresters for "preventing great numbers of good citizens from
acquiring the homesteads to which they were justly entitled." The roust-

about's remedy for this problem? He "expressed the idea that the college boys ought to be killed."[34]

In the larger arena, advocates of a new age in planning had to grapple with the frontier myth to create a usable past. Frederick Jackson Turner had set the tone by declaring, in 1893, that the encounter with wilderness had defined a peculiarly American character: resilient, resourceful, practical, inventive. Echoing Tocqueville, Turner noted that backwoods Americans were heedless of tradition or authority when common sense would serve just as well. The continual encounter between "savagery and civilization," Turner said, had shaped the "perennial rebirth" and "fluidity of American life." The "demand for land and the love of wilderness freedom" had pushed the frontier ever westward.[35] The problem, of course, was that the Census Bureau had declared the frontier closed in 1890. What, now, was to be the engine driving American achievement?

Turner, like many Americans, viewed the loss of the frontier in fretful and largely nostalgic terms. Without the hardening experience of constant rebirth, the nation seemed fated to slide into decadence — or worse. Even as Turner wrote, many thinkers were pointing to frightening changes in American society wrought by industrialization. These included loss of individual autonomy, concentrations of capital, and the consolidation of what, up to then, had been a relatively atomized economy. For many, the end of the frontier seemed to presage an unappealing choice between soul-deadening socialism or monopoly capitalism.

Others, such as Theodore Roosevelt, saw the frontier experience as a model for a new Darwinian struggle that would allow the best and brightest leaders to take charge of an emerging managerial society.[36] The novelist Owen Wister, a blue-blooded Philadelphian and member of TR's inner circle, saw the West as a breeding ground for a natural aristocracy of men — a place where the "quality" could rise above the "equality," just as the Founding Fathers had intended.[37] For Roosevelt and his friends, it was sufficient to keep the wilderness experience alive through occasional play-acting, such as Teddy's famous sojourns among the cowboys in Dakota. The novelist Frank Norris proclaimed that the vanished frontier should not be mourned at all. What counted, he said, was not the land itself but the Anglo-Saxon urge to conquer. For twentieth-century pioneers, the new frontier lay in worldwide commerce.[38]

The deepest thinkers of the World War I era rejected the frontier myth entirely, or at least declared it obsolete. This was particularly true among the young intellectuals who founded the *New Republic* in 1914. Like TR, they saw a strong managerial state as the key to a "middle way" between the equally undesirable poles of monopoly tyranny and socialist revolution. But they rejected the old Progressives' moral bluster in favor of a cool technocratic approach, one that would facilitate the guidance of society by enlightened experts. It was an ambitious and somewhat ironic scheme, for these new technocrats believed that only organization could preserve a playing field where individualism could thrive. In *The Promise of American Life* (1909), Herbert Croly had spurned the pioneer ideal, calling it unsuited to modern times. It celebrated mediocrity and led to a "worship of the average." In an age of interdependence and specialization, he wrote, the notion of pure self-reliance was a dangerous delusion.[39] Croly's colleague Walter Weyl was far more acerbic, declaring in 1912 that frontier individualism had psychologically "twisted" Americans to the point where they were incapable of uniting for the common good.[40]

The problem with the forest, of course, was that it *looked* like a frontier; thus the frontier myth was more intractable in the woods than anywhere else. Even as the forest was remade through the centralized agency of a managerial society, it necessarily retained qualities of the wild. The most visionary of the writers and planners of the 1920s would recognize this tension acutely. They would invoke the frontier myth—sometimes disingenuously—when it served the ends of organization. In the spirit of technocratic Progressivism, they would use organization to further individualism, but they would not tolerate the sort of ruinous dissipation wrought by total freedom. Still, the specter of the Turnerian frontier could not be entirely controlled. Indeed, as the 1920s progressed, it would express itself in ways that the planners could scarcely foresee.

CHAPTER TWO

The Agricultural Problem and
the Land-Economics Solution

[O]ver 100,000 farms of 80 acres each are still waiting for the settler
in Upper Wisconsin.
—HARRY L. RUSSELL, "Farms Follow Stumps," 1921

The problem which we have before us in land settlement is a
problem of democracy; namely, the problem of utilizing the brains
of the few who have capacity in the service of the many.
—RICHARD T. ELY, 1922

For American farmers, the "return to normalcy" meant a sudden slide
into economic depression. Exhorted by the Food Administration under
Herbert Hoover, they had tripled their exports during the world war,
only to see prices collapse when the wartime emergency passed.[1] In the
spring of 1920, agriculture hit the skids. By the summer of 1921, crop
prices had fallen by half. Prices would recover gradually after that, but the
damage was done.[2] At the end of 1923, the U.S. Department of Agriculture (USDA) reported that in fifteen hard-hit wheat- and corn-farming
states, an average of 8.5 percent of the farms had failed during that year.
An additional 15 percent were worthless on paper and were hanging on
only through the patience of creditors.[3] From 1920 to 1925, the U.S. farm
population would fall by almost two million people. And for the first
time in recorded history, the number of acres under cultivation during
that period would actually decline. Farms were modernizing at a break-
neck pace, making capital investments in tractors and hybrid seeds, yet
the rush toward "scientific agriculture" seemed to do little good for the

individual plowman. With scant ability to organize, and the relative importance of their industry in steady decline, farmers seemed to be caught in a cycle of despair.[4] What was worse, the lure of the city appeared to be robbing the farms of their best and brightest people. While government experts had fretted about a rural "brain drain" since the Progressive Era, now they became doubly concerned over the prospects for "cultural and racial decay" in the countryside.[5]

All of these woes were particularly acute in the Great Lakes cutover. Forty million northern acres in Michigan, Wisconsin, and Minnesota had been stripped of their choicest timber by 1900. Some of the lumber companies held on to their denuded lands for speculative purposes, while others sold them to land jobbers or simply abandoned them. With the encouragement of the states, an entire industry sprang up to sell cutover tracts to would-be farmers, many of whom were immigrants. It was widely supposed that hardy pioneers would pour into these lands to establish farmsteads. The cutover frontier might even provide a "safety valve" in times of social strife. Hotheaded foreigners, it was thought, could work out their frustrations by wrestling with stumps rather than turning to Bolshevism in the cities. Agricultural colleges lent a hand through extension work, confident that science could make the cutover prosper. The local press — fueled by advertising from cutover land-sellers — proclaimed a glorious agricultural future for the North. But cutover farmers were especially vulnerable to any turbulence in the agricultural economy. Even in good times, they had to deal with poor soil, a short growing season, and a lack of credit. Having made only marginal progress since the turn of the century, the push for cutover farming sputtered and stalled by the mid-1920s.[6]

From a distance of several decades, it is difficult to realize that the cutover crisis was essentially one of agriculture. Farms are largely absent from this land today, having been replaced by a "crop" of trees. Yet a thread of continuity runs through the cutover story, whether the crop involved was corn, cattle, or pine.

From the early days of settlement, state (later federal) authorities took a special interest in promoting prosperity by disseminating expertise among the populace. For agronomists, this meant doing basic and applied research, then sending extension workers into the field to spread the gospel of scientific agriculture among unlettered farmers. Often, too, this meant pushing farmers into "modern" methods — producing crops for market rather than just subsistence, buying fertilizer and machinery, and taking on debt for expansion. Harry L. Russell, who became dean of agriculture at the University of Wisconsin in 1907, was acutely concerned with bringing science and progress to the "under fellow" — the backward farmer who was suspicious of city ways.[7] The work of Russell and his colleagues seems to have consisted of equal measures of scientific faith, Progressive zeal, and friendly paternalism.

When the farm depression hit the cutover in the 1920s, a new generation of experts came to the fore. Prime among these were the land economists. Like their colleagues in agriculture, land economists were concerned with bringing expertise to bear on cutover questions. But land economics placed greater faith in centralized planning than in the ability of a rural populace to make good with a little expert guidance. Land economists sought to impart order and stability to the cutover, and to make it continuously productive. In doing so, they had to grapple with the mythology of individualism — a mythology that seemed well-suited to a rugged and desolate land that awaited the actions of individual pioneers to bring it to fruition.

Intellectually, the land economists were part of a small but influential American clique that spurned the idea of frontier individualism. To borrow Walter Lippmann's terminology, they favored the "mastery" of coordinated development rather than the "drift" of trial-and-error by ignorant individuals. Wisconsin's Richard T. Ely, a founder of the discipline, believed that only coordination by "men of character and brains" could avert "catastrophe" in the North.[8] Beginning about 1920, the land economists advanced the idea that much of the cutover should be off-limits to farming. At their best, they helped cutover pioneers avoid the heartbreak of trying to make a living in a region that often would not support it under modern standards. At their worst, they displayed an elitist, even vicious paternalism and an abhorrence of individual toil at the margins of agriculture. The methods they devised to remake the cutover — including the use of the police power to prevent settlement on "submarginal" lands — are largely responsible for the way the region looks today.

The Push for Farming

Boosterism and land promotion in some Great Lakes forest regions began as early as the 1860s. Minnesota, in 1870, assured would-be settlers that its lands could be tamed by pioneers "whose capital consists of brawny arms and brave hearts."[9] In Wisconsin, state geologist Increase Lapham predicted that forest settlers could make a steady living by selling provisions to workers in lumber camps.[10] In 1895, William Henry, dean of the University of Wisconsin's College of Agriculture, authored a hugely popular little tome for potential settlers, *Northern Wisconsin: A Handbook for the Homeseeker.* About fifty thousand copies had been distributed by state immigration officials by 1897. Henry saw the emerging cutover landscape as a solution to urban ills and the perils of Great Plains farming. In the lush northern woodlands, he predicted, city folks and European immigrants would find a welcoming home, much more hospitable than the plains of Nebraska or the Dakotas. The book's success was in part due to Henry's folksy tone; in its pages he assumed the demeanor of a wise and benevolent uncle. The land, Henry stated, would yield to human effort. Success or failure lay with the individual, though the state would help out where it could. Once cleared of trees, the North would not revert to "wilderness," Henry assured his readers.[11]

In tandem with state efforts, private interests hatched commercial colonization schemes. In 1902, for example, the Blue Grass Land Company of Minneapolis set up a "Finnish colony" of 25,000 acres in Vilas County, Wisconsin, on the border with Upper Michigan. It was thought that the region's forests and lakes, being similar to Finland in appearance, would prove pleasing to settlers. The company offered to "award 40 acres free of charge to the person who suggests the most appropriate Finnish name for the colony, one which at the same time is easy to pronounce in English." The settlement was named Toivola (Vale of Hope). The name proved illusory, for the colony—like most of its kind—soon failed.[12]

Writers and editors played a key part in all these enterprises. The most skillful of them spun vivid word-pictures of the region. Before about 1890, the abundance of free (under the Homestead Act) or cheap land in the West had averted many settlers' eyes from the cutover—why would a farmer pull stumps if he didn't have to? But with the easy land gone, cutover publicists began the task of redefining a region for an audience that often had never seen it. Through a sort of discursive alchemy, the cutover's drawbacks had to be turned into assets. In promotional literature, the region's deep winter snows would be likened to a "blanket" that insulated the earth and kept it moist for spring planting. The frigid northern climate was deemed "invigorating" and "healthful." And cutover promoters loved to write ominously of the droughts and plagues of locusts that surely awaited farmers who moved to the Plains. By comparison, stumps were minor irritants. Writers even redefined the cutover region through cycles of agricultural trial and failure. Boosters of Michigan's Upper Peninsula, for example, went through an enthusiasm for ranching (complete with cowboys in western garb) before settling on a pastoral dairying paradigm under the mellifluous name of "Cloverland."[13]

Pitching land to would-be farmers, cutover promoters relied heavily on pioneer themes. The cheap and stumpy acres of the North represented a chance to "make a home," prospective settlers were told. Success or failure depended entirely on the individual. If a man profited, he could expect every dime to go into his own pockets. A pamphlet for the Soo Line Railroad, circa 1915, invited the land-hungry settler to investigate northern Wisconsin, "where the factory whistle has no sinister meaning, where the pay envelope is exactly the returns of his own labor and brains." The Soo Line publicist also employed verse, a practice not uncommon in cutover literature of the era:

> He who owns a home of his own,
> If only a cottage with vines o'ergrown,
> Of the pleasures of life, gets a greater per cent,
> Than his haughtiest neighbor who has to pay rent.[14]

But the most concerted, coherent, and long-lasting push for cutover farming would come from land-grant colleges and their agricultural agents. The experience of one pioneering cutover agent, Ernest L. Luther of the University of Wisconsin, is instructive as to both the potential and the limits of "expert" advice in the Great Lakes forest regions.

On February 7, 1912, Luther arrived in Oneida County, Wisconsin, to begin work as the state's first agricultural extension agent. A recent University of Wisconsin graduate, Luther represented a bold experiment by the UW's College of Agriculture to bring scientific knowledge to cutover farmers. "I can see my job," he wrote as his train puffed northward. "Those *durned* stumps." Huge and deep-rooted, millions of pine stumps had to be pulled, burned, or blasted away before the cutover could be farmed. Luther, like many, viewed the stumps as a symbol of the cutover challenge: through sheer effort, the obstacles could be re-

Agricultural agent Ernest L. Luther in Oneida County, Wisconsin, about 1914. Luther believed that scientific expertise could transform the cutover into a prosperous farming region. Photograph courtesy of University of Wisconsin–Madison Archives.

moved and the way cleared for progress. "I have faith in the country," he wrote to his mentor, K. L. Hatch, at Madison. "We'll make good."[15]

In the days of unabashed agricultural promotion, agents such as Luther were part of a close-knit triad with local editors and land-sellers. In Oneida County, Luther was a good friend of the Brown family, publishers of the local *Rhinelander News*. The Browns also owned the county's biggest timber and land business, and were casting about for ways to ensure the county's prosperity (and their own) once the trees were gone. Luther often wrote for the *News* himself, urging farmers to build silos and to plant alfalfa. (So enamored was Luther of the UW's cutover prescription of corn, alfalfa, and silos that he referred to the formula as "our religious trinity.")[16] The Browns, for their part, were so enmeshed in the economic life of the community that it is difficult to tell where their community spirit left off and their self-interest began. It is clear that, like most others in Rhinelander, they genuinely believed that the plow would follow the ax. The family holdings included a farm, and whenever Luther contracted for rail cars of lime or fertilizer to enrich the thin soil of Oneida County, the Browns dutifully signed up for a ton or two. Even before Luther's arrival in Rhinelander, the *News* had boosted the region's agricultural potential, though not without dubious motives: "Every workingman in Rhinelander ought to be buying a piece of land while it is cheap. This condition is not going to last very long. . . . Every man can buy land on very reasonable terms, one quarter cash, balance in three years time, or in many cases, even more liberal than that." Not coincidentally, the bottom of that same page was occupied by an advertisement for "30,000 Acres of Farming and Stump Lands," for sale by the Brown Bros. Lumber Company.[17]

In time, though, some of the more enlightened promoters of agricultural settlement noticed a peculiar phenomenon: cutover settlers were failing, not because of any moral fault or laziness, but apparently because they lacked access to credit and long-term guidance. In the parlance of the time, they needed someone to "stake" them while they labored to put their farms on a paying basis. Benjamin F. Faast, who ran the Wisconsin Colonization Company, was well situated to recognize this fact. As a member of the University of Wisconsin Board of Regents, he had close contact with the agricultural scientists who were struggling to make the cutover bloom.[18] Colonization was a long-term effort, Faast believed; to succeed, a company had to offer ready-made farms in large, contiguous communities, backed by agricultural expertise and deep pools of capital. Cutover colonies needed their own private extension agents and resident bankers, he asserted. (Faast was careful, in this sense, to distinguish true "colonizers" from "get-rich-quick" promoters whose main goal was to sell land to suckers.) Even while clinging to an agricultural formula, Faast and his brethren represented a transition stage between pioneer independence and centralized planning. They still viewed the cutover problem largely as a matter of tending to the needs of individuals. But they also recognized that uncoordinated development, left to the whims of ignorant pioneers and the greed of land hustlers, would yield untold and unnecessary misery.[19] The UW's Richard Ely, who was somewhat sympathetic to well-planned colonization, believed that intelligent planning could put good farmers on good lands. But the "old individualism is gone," Ely told an audience of land-sellers and colonizers in 1918. "Those who try to cling to it in land settlement as well as otherwise will finally die with it."[20]

In the main, cutover farming failed dismally. Crop yields were skimpy, and credit often was available only at usurious rates, if at all. Coupled

A cutover homestead of the Wisconsin Colonization Company, probably about 1920. Photograph courtesy of State Historical Society of Wisconsin, WHi (X3) 52242.

with the high debt-to-equity ratio and lack of experience of many cutover farmers, it was a formula for ruin. On some lands in Michigan that were sold to unsuspecting buyers on the installment plan, nine of every ten farmers failed.[21] In northern Minnesota, the few choice farmsteads were greatly outnumbered by "sandy land, swampy land, and land that could yield little."[22] The last great boom of cutover enthusiasm was a suggestion from Interior Secretary Franklin K. Lane that returning soldiers from the world war be settled on cutover lands with federal sponsorship. The scheme was defeated by Congress, which feared that it would spawn an uncontrollable bureaucracy. At the same time, some academics were beginning to theorize that the United States had too much land in farms, not too little. Then the agricultural depression of the 1920s hit with a vengeance. Suddenly, the idea of sending doughboys onto the land seemed more a punishment than a reward. It never happened.[23] Instead, the 1920s would see the ascendance of newly sophisticated legal-economic strategies for the cutover. In places such as Oneida County, where every new farmer formerly had been regarded as a harbinger of progress, the upshot of these policies would be to discourage agricultural settlement — and eventually, to forbid it entirely.[24]

Laws of the Land

The farm crisis of the 1920s would capture the imagination of one of the most venerated academic figures of the time: Richard Theodore Ely, an economist at the University of Wisconsin. Urbane, famous, and by most accounts somewhat aloof, Ely was in his sixties when he began casting about for a new challenge. He found it in an unlikely place: the Great Lakes cutover. At an age when most men might retire, he would help to invent the new discipline of land economics.

Educated in Germany, Ely had sealed his reputation by questioning the assumptions of classical economics, including the assertion that economic laws were universal and unchanging. For Ely, economics did not reside in dusty books, but in the world at large.[25] Grounding his work in empiricism rather than mathematics, Ely believed that economic remedies should change with the times and that economic reasoning should be "inductive"; that is, it should start with the facts.[26] Like his student and colleague John Commons, Ely asserted that social problems could be addressed by pragmatic economic analysis aimed at reform.[27] His outlook would turn decidedly conservative in his later years, but Ely never abandoned his vision of an activist state and a utopia of expertise.

In 1920, Ely founded the Institute for Research in Land Economics and Public Utilities at the University of Wisconsin. The institute brought together diverse strains of thought on land-use questions that had been in development for a decade. Basically, Ely and a handful of others sought to apply an inductive method to questions of land use, both rural and urban. (As the institute's title suggests, they also studied questions of utility ownership, development, and regulation.) Because the institute's work was purposely worldly, it sought laboratories where data might be gathered and practical solutions tested. One of the first such laboratories was northern Wisconsin.[28]

Before about 1920, the land economists shared much of the agriculturalists' faith in cutover farming, but only under terms of settlement that were tightly controlled. In time they would conclude that much of the cutover was best suited to growing trees. Throughout this evolution, the land economists stressed the need for what Ely called "economic democracy" — a social equation that balanced individual opportunity

Richard T. Ely, who employed the Wisconsin cutover as a laboratory for the new discipline of land economics. Photograph courtesy of State Historical Society of Wisconsin, WHi (X3) 17722.

with societal goals.[29] Preservation of freedom was a paramount aim, but to achieve it, the actions of individuals would have to be channeled to curb waste and to limit the possibilities of social unrest. The frontier was gone, and with it the days of boom-and-bust. Neither individual profligacy nor corporate treachery could be sanctioned. Land economics was a social science, in that it aimed to improve "the social conditions of living."[30]

His mission was far-reaching but not without precedent. In 1912 he had given a series of informal talks on the breakdown of laissez-faire in Europe. The Germans, he asserted, had negotiated a delicate middle way between the equally unappealing poles of socialism, on one hand, and a despotic collusion between entrenched privilege and monopoly capital on the other. A "theory of regulated economic relations" — including forest conservation and public ownership of utilities — was helping to preserve "general prosperity," "liberty," and the "maintenance of private property and individuality." It was "the opposite of socialism," in that it "aims to maintain the existence of the present socio-economic order in its essential elements." Americans, Ely believed, should follow suit. Indeed, if they hoped to ensure stability and prosperity while preserving a semblance of individual opportunity, they had no choice.[31]

How would this conception be applied to the cutover? Fundamentally, land use had to be consciously shaped and directed from above. The goal was to make the land continuously productive by reducing waste and assuring each man a chance to reap a fair reward for his labor. The land economists' vision of rural life stressed several points:

Independent ownership of farm land. The early land economists were acutely concerned with "tenancy" — the renting of land, provision of farm homes in return for labor, or sharecropping. The practice was widespread in the South and was becoming popular in the Great Plains. Ely and his colleagues feared that it would spread into the cutover, where settlers often lacked capital and were at the mercy of land-sellers. To ensure social stability, lands should be "cultivated in the main by their owners," Ely believed.[32] While some tenancy was unavoidable, planners should ensure that the "agricultural ladder" afforded tenants the opportunity to become independent farmers if they were capable of it. (With typical indelicacy, Ely stated that tenants of lesser ability should be "weeded out" of the farm economy.)[33] Ironically, the land economists hoped to use centralized coordination to create a landscape that was almost Jeffersonian in character.

An abhorrence of waste. The Progressives, of course, had made "efficiency" a byword. But like many such terms it was overused and ill-defined. The academics, social scientists, and policy makers of the ensuing years would work to refine it. Frederick W. Taylor had set the tone with his work in scientific management. Through centralized study and direction, the labor of individuals could yield greater rewards for all. By coordinating their workers' labor, factories could increase productivity so much that they could double their wages and effectively buy labor peace, Taylor asserted.[34]

The most disturbing—and most preventable—form of waste in agriculture was the application of capital and labor inputs to land that would not reward those inputs sufficiently. Put simply, a farmer would sweat just as much trying to cultivate bad soil as good. If he failed, his labor, his time, and his good intentions had been wasted, and society as a whole was less stable and less prosperous for it. "A crop which will not pay for itself with a fair profit is not worth putting into the ground, and the soil should be put to some other use which will pay for itself," Ely believed.[35] In frontier days, pioneers had discovered good land through trial and error. Now the government could lead them to it.

A near-worship of aggregate information. Agronomists had long been interested in soil types, but the land economists sought wider and more conclusive data concerning land use. How much of a given crop could a farmer expect from a given plot of land? Would he have access to credit and to transportation facilities? What were the relative returns of a given acre in different uses—for farm crops, pasture, or growing timber? Which acres would reward the farmer's investment most bountifully? As always, the inductive method applied: "Facts are sought first and from these can be discovered the trends or principles which will be useful as guides to action."[36]

A recognition of the "social side of private property." The frontier era, when the economy was "managed" primarily through millions of atomized private decisions, was gone.[37] Now, a new generation of planners would assess the social costs of private decisions. Through legal-economic measures, policy makers would reward desired private behavior and punish the undesirable—even prohibit it altogether. Ely predicted that more citizens would be "forced" to recognize their duty as time wore on.[38]

The rating of human beings along with acres. Ely believed that people as well as land could be ranked as to their fitness for farming. Besides the good farmers, there were "marginal" and "submarginal" ones. "It goes without saying that if no submarginal land were in use by farmers and if there were no submarginal farmers, farmers would be prosperous," he declared at one point.[39] The drive for classifying men, Ely believed, reflected a larger natural order, which had decreed that some people were more equal than others. While he rejoiced in American opportunities, he was offended by the myth that anyone could rise above the common herd at any time: "All our stories, all our teaching and preaching tend to dangle before the eyes of the poorest and meanest the few great prizes of life." The duty of the elite, he asserted, was not to encourage every "rail splitter" to become president. Barring an occasional Lincoln, it would not happen. Rather, the enlightened few should help the masses of humanity to discover their "proper and natural relations."[40]

The Mission and Its Meaning

By the early 1920s, statistics told the cutover story in stark terms. Much of what was occurring in Ely's northern Wisconsin "laboratory" also was happening elsewhere. As lumbermen picked through the remaining forests in the Great Lakes states, they continued to cut their vast acreage in the American South and the Pacific Northwest. As a result, the amount of cutover land in the United States was increasing by 16,600 acres — almost twenty-six square miles — every day.[41] Much of that land would not support farms under a modern standard of living. And agriculture as a whole was reeling after the crop slump of 1920–21. In the previous twenty years, the United States had gone on a binge of land-clearing and reclamation — carving farmsteads from the woods, draining swamps, and irrigating deserts. Now, instead of much-feared food shortages, farmers were producing chronic surpluses. In some areas, particularly New England, farms already were being abandoned and were reverting to woods. After World War I, some agriculturalists and land economists began suggesting that the nation had too much land in farms. The result of that imbalance, they asserted, was wasted effort on substandard acres and diversion of that land from more socially desirable uses. Prime among those uses was the growing of timber.

In the Great Lakes states, the cutover crisis engendered deep conflict among downstate "experts" as to the proper role of colleges and governments in agriculture. The assertion of "too many farms" was not met with universal joy among agriculturalists. Many of them, such as the UW's Harry Russell, had devoted their lives to helping struggling farmers succeed at the margins. As early as 1919, his well-meaning boosterism had clashed with Ely's gloomier view of cutover prospects.[42] The sharp contrast between the two men was personal as well as professional. Russell was mindful of farmers' influence with the legislature and their lingering distrust of "book farming" and was acutely conscious of the university's need to get along with individual plowmen. Despite his exceptional intellect, he conceived of himself as a sort of agricultural father figure, helping yeomen to help themselves. Ely viewed agriculture from a safer distance. Many of his ideas were hatched not in the field, but in his elegant home in Madison, which doubled as an academic salon. His concern for the future of agriculture was mixed with a certain contempt for individual farmers. As the cutover crisis worsened, he justified his increasing paternalism by reasoning that cutover farmers were not especially bright. If they were, he figured, they wouldn't have tried cutover farming in the first place.[43]

After 1920, agricultural leadership in the United States underwent a fundamental change. The old ethos had stressed research and dissemination of technical expertise to help farmers grow crops. The new ethos augmented the old with an overlay of social-science methodology, aimed at advancing a broader conception of the public good. Its work encompassed disciplines such as land economics and its subspecialty of land utilization, rural sociology, and agricultural economics. It abandoned the idea, long cherished by many agronomists, that farming was a societal good in itself. The shift was signaled most dramatically by the issuance of the 1923 *USDA Yearbook,* which contained a pathbreaking report on land utilization. O. E. Baker and L. C. Gray, both of whom were disciples of Ely, declared in the report that there was no danger of a dearth of farmland and that cutover acreage should be shifted to forestry to deal with the worsening timber shortage. Within a few years, this broadened conception of the land problem would result in a slowdown in federal reclamation projects and an increase in the monies allotted to forestry.[44]

Like Ely, many of the new experts had enjoyed a comparatively rarefied upbringing, with upper-middle-class origins and an emphasis on higher

Harry L. Russell, the brilliant and headstrong dean of agriculture at the University of Wisconsin. A self-described friend of the farmer, Russell pushed for cutover homesteading well into the 1920s. Photograph courtesy of University of Wisconsin–Madison Archives.

education and public service. One historian says these factors often led to a "feeling of detachment" and a sense of divine mission. Henry C. Taylor, the father of agricultural economics, had studied under Ely at Wisconsin, where he absorbed the master's worldview and, apparently, a measure of his sizable ego as well. At one point, asking churches to help fund graduate students in the rural social sciences, Taylor opined that "these men and women who are benefiting by special educational privileges are the ones who . . . can do the most effective work in the establishing of the Kingdom of God on earth."[45]

This attitude of righteous detachment sometimes led researchers to regard individual farmers as mere guinea pigs in a glorious agricultural experiment. Rural sociologist Charles Josiah Galpin, who led the USDA's research into the quality of farm life during the 1920s, reflected many researchers' concerns about the apparent flight of "desirable" people from the countryside. Galpin argued that farm policies should be aimed at producing "parity" in agriculture: the efforts of an intelligent man should yield as much reward in the country as in the city. Only this would keep the "modern cerebral" type of farmer on the land. Otherwise, farm life would come to be dominated by "primitive muscular types" who would drag down the quality of rural living and ultimately degrade the rural gene pool.[46] Through the 1920s and into the 1930s, many cutover planners would reflect these strains of thinking—through an abhorrence of individualism, scorn for the pioneer myth, or even a subtle advocacy of eugenics. The friendly paternalism of earlier days would take on a much meaner edge as the agricultural crisis deepened. "In the minds of professors and other experts," one historian writes, "cutover settlers were changing from clients to be assisted . . . to victims to be protected . . . to waste products to be feared."[47]

As the 1920s wore on, the theorists and their partners in policy making devised numerous legal-economic methods to reshape the land. Measures such as tax incentives, fire protection, and rural zoning would help remake the cutover in the image Ely desired. But with each step, the members of the cutover brain trust found that they had to explain themselves to various publics—fellow academics, legislators, the "booster" press, skeptical farmers, and finally the public at large. Some of their desires could be had only with the assent of the voters. To gain that assent, they had to ground their ideas in real-world examples and counter some prominent American myths—especially the lingering visage of the frontier. From his lofty perch at Madison, Richard Ely could hardly realize that the cutover revival would be bound up as tightly with popular ideals and aspirations as it was with the work of social scientists. Others saw opportunity where he did not.

CHAPTER THREE

P. S. Lovejoy and the Campaign for Order

What [the cutover] needs is a receiver in bankruptcy.... The
Cloverland trouble is not so much like a railroad wreck where there
is a physical smash-up as like a railroad wreck in the manner of Wall
Street, where the smash-up is financial and moral.

— P. S. LOVEJOY, 1920

The chief irony of the American frontier is that it appears to have cap-
tured the popular imagination even as it was fading from actual memory. The interrelated strains of the frontier myth — the celebration of
the backwoodsman and trapper, the glorification of the Indian-fighter,
and the enshrinement of the yeoman farmer — grew more forceful, not
less, as the nation became urbanized and its people became more interdependent.[1] Indeed, Richard Hofstadter has written that the "agrarian
myth" of landed independence came to be "believed more widely and
tenaciously as it became more fictional."[2]

Yet the glorification of the pioneer was not simply an exercise in nostalgia or antimodernism. At its core lay a uniquely American ethos of
optimism and anticipation, one that had been noted early on by observers
such as Crèvecouer and Tocqueville. Even Frederick Jackson Turner, who
was hardly sanguine about the prospects of America without a frontier,
had written in 1903 that the log cabins of the forests and prairies were
the natural forerunners of "the lofty buildings of great cities." For Turner
and others, the concept of hewing civilization from wilderness was a
symbol of nation building, rather than of rusticism for its own sake.[3]

For the planners of the 1920s, the key task was to pay obeisance to the
pioneer myth even while seeking to transmute it. If Americans clung to

ideas of progress, faith in the future, and individual fulfillment, the mission of planning advocates was to advertise collective action as a guarantor of individual aspirations. In this sense, the writers who "reimagined" the American landscape in the 1920s would have to define a new frontier of their own: the idea that the best friend of the individual could be the state.

In the summer of 1919, Michigan's Upper Peninsula welcomed an odd visitor: a frail and bookish-looking man from the university at Ann Arbor, an urban specimen who nonetheless talked of Paul Bunyan and loved to play the banjo. For several weeks, Professor Parish Storrs Lovejoy was squired around the region like visiting royalty.

Upper Michigan had been stripped of its best timber, but Lovejoy's hosts now sensed that something better was growing. Bumping down the logging roads in Model Ts, the local boosters treated Lovejoy to a display of emerging agricultural splendor. "They showed me silver linings and sweet clover, . . . Finns and Polanders with their women barefoot in the fields, . . . towns coming up out of the wrecks of logging camps, . . . all manner of interesting things," Lovejoy recalled.[4]

But the tour was hardly representative of the larger country, and Lovejoy knew it. As he journeyed through the former North Woods on his own, Lovejoy found mostly "brushy wastes of scrub, fire-weed forests, bleached snags and charred stumps." The boosters' dreams for "Cloverland," as they optimistically called it, clearly were not coming true.[5]

Agricultural progress in the region had been slow. Lovejoy found a handful of settlers trying to grub a living from sandy, acidic soil. Pockets of rich land were interspersed with poor, but the settlers had tried to farm the bad soil as well as the good. Land speculators had managed to unload just a fraction of the former pinelands, the rest sitting idle until its "carrying costs" — interest on mortgage debt and annual property taxes — exceeded any foreseeable gain from selling it in the future. The record of their failure was written in the tax rolls. Vast areas of "Cloverland" had simply been abandoned, reverting to government ownership for nonpayment of property taxes. "Millions of acres of it have reverted for taxes and have been peddled out again for pennies an acre," Lovejoy noted. In many cases, land valuations were so low that counties spent more to collect the taxes than they received in revenue. The slick patter of the "land boomers" would not alter a central fact: "A good part

P. S. Lovejoy. Photograph courtesy of State Archives of Michigan.

of Cloverland is starvation poor, because the soil naturally is nothing but lean and hungry sand."[6]

Beginning that summer and for the next several years, Lovejoy unleashed a barrage of magazine articles on the subject of northern development. The cutover region had a glut of land, he wrote, and no amount of scheming would fill that land with farmers. On much of this land, agriculture simply didn't pay. Soils were poor, growing seasons were too short, and credit was hard to come by. In the early 1920s, even the most prosperous American farmers were haunted by persistent surpluses and falling prices for their products. Against that backdrop, sending an unknowing settler into an unforgiving region like the cutover seemed unwise — even cruel. Government agencies and private firms, Lovejoy declared, had no business promoting settlement in areas where farmers were likely to fail. And farmers, he came to believe, had no business being on land where their efforts would yield only a substandard mode of living. "Nearly everybody has always taken it for granted that the more land there is in farms and the more farmers there are on the land, the better off we all are," he wrote. But the disappointments of cutover farming had proved this maxim wrong. States now had a duty to inventory their rural lands and to classify them according to best use: for farming, grazing, recreation, or the growing of timber as a crop. The goal was to make all acres productive in some form, but to discourage farming on the millions of acres that would not support it under modern standards. "The old land-booming days are just about over," Lovejoy declared. "Land-booming methods have failed and worse than failed. Something new and different and better is coming up."[7] The scruffy and burned landscape of the cutover, he believed, must give way to one characterized by economic rationality, expert direction, stability, and order.

P. S. Lovejoy, perhaps more than any other person in the United States, was equipped both to formulate such a vision and to sell it to the masses. In the small world of 1920s conservation, Lovejoy was a genuine rarity: a trained scientist with a flair for the backwoods vernacular. Schooled in the German forestry tradition at the University of Michigan, he had worked for Gifford Pinchot as a forest supervisor in Wyoming and Washington state, then signed on as a professor at the Forestry School in Ann Arbor in 1912. But the bloodless atmosphere of academe left him restless and dissatisfied. In 1920 he quit the Michigan faculty, embarking on a three-year odyssey of bold and prolific free-lance writing.

Most of his work appeared in the *Country Gentleman,* a weekly farm magazine owned by the powerful Curtis Publishing Company of Philadelphia. In several dozen articles, Lovejoy hammered away at the need for land-use planning, replanting of cutover areas with a new "crop" of trees, and regulation of real-estate sharks in the woods. Each article was an engaging lesson in land economics, delivered in the folksy tones of a cracker-barrel scholar. But Lovejoy's easy manner veiled a sharp and tireless intelligence, one that was pushing the boundaries of both popular and academic thought on land-use questions.[8]

In retrospect, his freelance work of the early 1920s amounts to nothing less than a total refiguring of the social contract under which Americans had subdued, reshaped, and employed the American landscape. Rugged individualism had worked well to open up the Ohio Valley and other areas suitable for agriculture. But Lovejoy recognized, earlier than most, that the cutover would not yield so readily to individual effort. He also realized that the cutover polemicists had to forge a clear economic argument for their work. Lovejoy and his colleagues shared a vision of rational land use, and they invented the legal measures and the bureaucratic apparatus that made that vision real. At the same time, they were coining a new cultural vocabulary that would make sense of their work for a larger audience.

The *Country Gentleman* made a good venue for Lovejoy's campaign. Its owner also published the *Saturday Evening Post* and the *Ladies' Home Journal.* The company's pockets were deep. Lovejoy received $350 for each feature article — a sum that allowed him to travel as he wished, while making a better living than he had as a college professor. The magazine circulated about eight hundred thousand copies a week in a nation whose principal occupation still was farming. Lovejoy's editor, John E. Pickett, shared Lovejoy's passion for forestry and urged the author to attack the issue head-on. Some of Lovejoy's peers questioned his move away from academia and into journalism. But Lovejoy believed that questions of land use — particularly the reforestation of the cutover — depended heavily on popular support. Freelancing, he told one forester, "has gotten me into contact with entirely new men and agencies through which to sell the timber-crop idea" and "has caught me a fine set of enemies that I am proud of." He was confident that he was reaching "a new audience and a big one."[9]

An agricultural audience was a logical choice, because the fate of the cutover was largely an agricultural question. Land speculators, farmers, local boosters, extension agents, and agricultural-college scientists tended to assume that the former North Woods would give way to farms. Lovejoy found that prospect unlikely, even unwise, arguing instead that a "crop" of new trees was the region's best hope for long-term stability. In this respect, the *Country Gentleman* was an ideal forum for what Lovejoy called "banana-peel engineering" — the placement of key ideas in places where people were likely to slip on them.

The Pioneer Land Bargain

Lovejoy's challenge was daunting: he had to employ scientific rationale to debunk popular thought about land use dating from America's pioneer days. Interwoven with the mythology of individualism, these beliefs were deeply ingrained in the national psyche. The old "land ethos" consisted of three elements:

   1. *A belief in the efficacy of widespread land dispersal and private, atomized decision-making.* Nineteenth-century governments had used land as an economic tool for mobilization of capital and labor, both of which were in chronically short supply in most of the United States. Hundreds of millions of acres in the federal domain had essentially been given away (the price ranging from nothing to a dollar or two an acre) in hopes that settlers would occupy the land. Federal, state, and local governments undergirded this effort with a matrix of laws designed to establish property rights and to encourage economic risk-taking. Farms gave rise to towns, which attracted newcomers from the more settled areas of the East. A continued belief in Jeffersonian ideals of yeomanry only encouraged the quick dispersal of the public lands.[10]

   2. *"Town building" as a means of economic uplift.* For the individual settler, the prime motivator — and, over time, the prime expectation — was the prospect of capital gains in land. The pioneer faced years of isolation and back-breaking labor, but he could count on his land rising in value as others joined him on the frontier. In time, schools, churches, and other amenities of settlement would spring up around him. The hardships of pioneer life thus were offset by the expectation of growing wealth and civility as communities expanded. A Minnesota promotional book issued in 1870 promised that the pioneer would soon be "not a

lone settler... but in daily communication with the busy world, and the proud possessor of a valuable farm which has cost him little but the sweat of his brow."[11] Economic historians refer to this phenomenon as "town building."[12]

In 1931, geographer Isaiah Bowman noted (with some dismay) that this characteristic American optimism over land values had persisted well into the twentieth century. Bowman acknowledged the innate attractiveness of the "town building" ideal and, in describing it, seemed to have succumbed to some of its charms himself. The pioneer, he wrote, "had poorer schools and roads, less social life, a cheaper house, indeed oftener not a house at all but a hut. But he had land that was his own, and he had enough of it for a living. The comforts could wait, and he could be cheerful because he knew that he could secure them in time. A well-defined cycle of benefits, in the period between 1840 and 1890 in the Middle West country, was widely expected. Land was bound to increase in value."[13]

3. *A continued official belief in the advisability of "incremental" farming.* The closing of the frontier, a fact recognized by the government in the early 1890s, spawned numerous anxieties, not the least of which was the fear that the United States would not be able to feed itself. This anxiety lingered into the 1920s. Agricultural scientists and extension agents thus committed themselves unconditionally to helping farmers grow more crops. The "technical agriculturalists," as Lovejoy dubbed them, thought not in terms of whether farming was wise in a given situation, but simply whether it was possible. Every farmer added at the margin of settlement was regarded as a net gain for the larger society. In 1916, for example, the U.S. Department of Agriculture reported on a study of 801 farms in the cutover districts of northern Michigan, Wisconsin, and Minnesota. The federal researchers found that many of these farms were operating just a notch above subsistence level. Nonetheless, they endorsed cutover agriculture as a sound proposition: "From a strictly business point of view these farms do not appear to be successful, but they furnish a home for the family and offer an opportunity to earn a living." The College of Agriculture at the University of Wisconsin emphasized land clearing in the cutover. Dynamite and, later, war-surplus TNT were touted as the best means for ridding the land of stubborn stumps.[14]

The local press in cutover towns was a key ally in the drive for agriculture. Northern editors resented suggestions that their lands would not support farming, and they railed against proposals for the planting of forest reserves. "This is not a barren wilderness," the *Rhinelander News* of northern Wisconsin declared in 1912. "We have as good soil and as fine a climate as any locality in the state."[15] Even in the face of the farm depression, this boosterism endured well into the 1920s. The agricultural colleges augmented this effort with a barrage of press releases and farmers' bulletins praising the agricultural potential of cutover lands.

A New Formula for the Cutover

P. S. Lovejoy was thirty-five years old when, in 1919, he made his first serious foray into magazine work. In a three-part series in the *Country Gentleman,* he introduced readers to the basic contours of the cutover situation. Millions of acres in Michigan, Wisconsin, and Minnesota stood idle, even as the nation faced the possibility of a timber shortage. The Great Lakes states, whose timber had built the settlements of the Plains, now had to import most of their wood products from the South and the Pacific Northwest. Whole counties in the former lumbering districts were "practically bankrupt," Lovejoy asserted. Contrary to boosters' assurances, the plow would not follow the ax. It was time to talk of reforestation.[16]

Even as the series emerged from the presses, Lovejoy was back in the North for his grand tour of "Cloverland." The trip was a whirlwind introduction to the booster mentality of the cutover region. John A. Doelle, secretary of the Upper Peninsula Development Bureau, escorted Lovejoy through the countryside and introduced him to extension agents, newspaper editors, businessmen, and settlers. Everywhere the talk was of agriculture. The lumbering days were remembered with a certain wistfulness, but the passing of the forest was not lamented. The developers of "Cloverland" were casting about for opportunity, seeking a magic formula that would make the region thrive. The latest idea was grazing. Cutover cowboys, adorned in appropriate hats, chaps, and spurs, roamed the stump-filled range that summer. The one crop that was never thought of was timber. Indeed, the forest seldom intruded on the minds of anyone in "Cloverland," even when its remnants caught fire and threatened to turn the entire region to cinders:

> The visit of an uncredentialed western ranchman set the whole district agog and clacking, but a 10,000-acre fire or a town packed into box-cars and ready to abandon homes to the ubiquitous smoke was accepted as a

mere temporary discomfort or nuisance. And kindly to observe our rutabagas. Are they not excellent rutabagas?

No agricultural fantasy seemed to be too extreme to be acceptable. My inquiry as to the prospects for a tomato canning factory on the shores of Lake Superior was seriously accepted and debated.[17]

The above paragraphs were written for *American Forestry*, whose readers presumably shared Lovejoy's views on the problems of cutover agriculture. In the *Country Gentleman*, Lovejoy's tone was far more circumspect. His years in academia had not blinded him to the social and political mores of the North Country. While a forestry student at Michigan, he had worked alongside timber crews in the fast-disappearing woods (Lovejoy had been measuring and otherwise studying the trees as the lumbermen cut them down). He understood the lumberjacks' disdain for "college boys" and for reforestation, which presumably would prevent honest homesteaders from establishing farms. Lovejoy realized, too, that "many farmers have followed their dads and grand-dads in hating the trees and the stumps which kept their plows from the good fields." Lovejoy knew he would have to acknowledge popular beliefs before offering something in their place. He wrote many of his cutover articles as running dialogues, often with himself in the role of innocent rube conversing with a skeptical reader. In one piece he extolled the virtues of "Cloverland" at great length until the exasperated reader told him, "Lay off the poetry and get down to it." At which point, of course, he did.[18]

Lovejoy was a skillful and patient polemicist. He never scolded his readers. Rather, he outlined the conventional wisdom about land-use questions — then systematically tore it apart. One article in the 1919 series, for example, included a picture of a young family arrayed in front of a rude log cabin in the woods. For many readers, the picture must have struck a chord of nostalgia and sympathy. Perhaps the frontier was not gone after all! Lovejoy recognized such yearning himself: "We are still close to the old pioneer days. Everybody takes a sympathetic interest in the new regions and, on occasion, helps out the fellows who have the nerve to pioneer it to-day. Didn't Lincoln live in a log cabin?"[19]

Indeed he had, Lovejoy assured his readers. He had lived in several of them. In one instance, the Lincoln family had spent fourteen years farming "niggardly" soil in Indiana before deciding that the situation was hopeless and moving to Illinois. Lovejoy checked a census report for that Indiana county and found that its population was estimated to have fallen by 8 percent since 1910. Apparently farmers there were still learning hard lessons, just as the Lincolns had. Wouldn't it be sensible, Lovejoy asked his audience, for government to lend a hand by identifying lands where farming would pay — and lands where it almost certainly would not?[20]

Despite the relatively gentle approach, Lovejoy's early cutover articles drew an avalanche of mail. A lumberman-turned-land-seller complained that Lovejoy must have been "blindfolded and had his ears stuffed with cotton, to write many of his statements." The man threatened to "tell on" Lovejoy to the dean of his state's agricultural college. A Minnesota real-estate operator called the articles "venomous and exaggerated." A chamber of commerce official said they were "calculated to do us tremendous damage." Yet, as if to suggest that attitudes in the North Country were vulnerable to persuasion, the mail was not entirely hostile. One reader enclosed a clipping telling how an Iowa man, who had bought a Michigan cutover farm in the dead of winter, had killed himself when he saw what the land looked like with the snow off.[21]

Lovejoy knew it was fruitless to argue that the majority of the cutover was suited only to growing trees. The idea of "town building" was too strongly ingrained at the local level; to tell local boosters that their land should revert to forest was to invite a fistfight. These boosters sometimes referred to the forest as a "jungle" and reforestation as "re-jungleization," as if it were something that flew entirely in the face of common sense. The prevalent attitude was summed up by Aldo Leopold: stumps, he noted, were considered to be a sign of progress.[22] The idea that a forest was desirable — economically, aesthetically, or otherwise — would have to be instilled gradually, and only after alternative uses for the land had foundered.

Instead of simply pushing forestry, then, Lovejoy embraced an alternate strategy: the idea of "service in land deals." Lovejoy stressed that land should be inventoried, and that governments should clearly identify the lands that were most likely to provide a decent living to farmers. It was a clever message, for Lovejoy was taking the farm community's traditional hostility to forestry and turning it on its head. If the "best" lands were identified, by extension the "worst" ones (those more suited to forestry) would be, too. No longer would the cutover farmer toil for years on unpromising soil. Instead, he would invest his time and energy in places most likely to repay the investment. Unless a plot of land could guarantee "a genuine and first-class chance for a good, safe living and reasonable profits," Lovejoy wrote, its development constituted "mere wild-catting and something to be stopped."[23]

In a 1921 article, Lovejoy applauded Wisconsin's state government, which had taken tentative steps to warn prospective settlers against land fraud. (This was a vivid contrast with other states, whose officials often assembled "sucker lists" of land seekers and provided them to real estate agents.) He also endorsed the idea of planned communities in the woods, whose developers would provide roads, farmhouses and outbuildings, and even livestock: "Ways must be found to let the pioneer settler get on his feet and going." Lovejoy probably doubted that many land-sellers would make such an investment of time and money—and in fact, few of them did. But he was astute enough to know that, while he built his case for forestry, he could not appear hostile to the farmer.[24]

Schools of Thought

Lovejoy's venture in journalism coincided with the last great push for cutover farming, a postwar campaign waged by agricultural colleges and state and local development bureaus. For college agronomists, the cutover was an irresistible challenge. An "atmosphere of glowing optimism" pervaded the agricultural labs of the University of Wisconsin in the early 1920s, one student of the period has written.[25] Extension agents and experiment stations focused special attention on the North, dealing with problems such as stump-pulling, selection of farm sites, and the search for hardy crop varieties. The problem, in Lovejoy's view, was that the agricultural colleges had failed to distinguish "between agricultural *possibilities* and agricultural *practicabilities*. The real issue is not one of farming technology but one of land economics. The agronomist and farmer 'might,' but *have* they?"[26]

The solution, as Lovejoy saw it, was not just individual initiative or boosterism (even if conducted in good faith, which Lovejoy assumed most of it was), but the collection and application of aggregate data regarding the cutover. Without the facts, land-use problems were certain "to snarl worse and worse," Lovejoy believed. He repeatedly commended the work of Wisconsin's Richard T. Ely, including his efforts to discourage settlement on "counterfeit," or substandard, crop land. "From now on," Lovejoy declared about 1922, "instead of merely guessing or lying

University of Wisconsin officials prepare to board the "Land Clearing Special," a promotional train that toured the cutover, about 1920. The train drew large crowds with its demonstrations of land-clearing tools, such as mechanical stump-pullers and war-surplus explosives. Photograph courtesy of University of Wisconsin–Madison Archives.

about it, we are due to get the facts as to the use of land and to base land use and land development on those facts."[27]

Until about 1925, thought regarding the cutover was seriously in flux. Land economists and agriculturalists often worked at odds. The confusion and conflict in academic circles was replicated elsewhere. One barometer of such confusion was the Tri-State Development Congress. Called by the governors of Michigan, Wisconsin, and Minnesota, the Congress met annually from 1921 to 1924 to discuss cutover development. C. L. Harrington, Wisconsin's chief forester, attended the meeting at Milwaukee in 1922. Most of its sessions, he told Lovejoy, "dealt with carloads of dynamite, 450,000 new settlers," and optimistic forecasts for land clearing. The role of forestry was acknowledged, but "no one has yet been found to pay the bill." Others were boosting the region as a tourist haven, the "Playground of the Middle West," "but there will soon be no playgrounds left, if everything is busted up with trainloads of dynamite," Harrington noted sardonically. A. D. Campbell, Wisconsin's former commissioner of immigration, assured Lovejoy that farmers could make a go of it in the cutover if they only got a boost from government, a "grubstake" that would tide them over until they cleared enough land to start making profits. Millions of federal dollars were being spent

A cutover stump yields to explosives. Some University of Wisconsin officials encouraged cutover land-clearing until the mid-1920s. Photograph courtesy of University of Wisconsin–Madison Archives.

on western reclamation projects; why not divert some of that cash to cutover settlers? Lovejoy's blunt dismissal of the scheme seemed to indicate his growing exasperation for those who assumed cutover prosperity was just around the corner: "If the lands are 'supreme' and ... there is a widespread desire on the part of small farmers to acquire land for farming, just why should it be necessary to involve federal tax money in developing these lands?"[28]

Lovejoy defined conservation as "reason applied to environment." His arguments for reforestation relied on methodical, rational explication of the cutover problem and its possible solutions. The strategy was deliberate. Lovejoy recognized that conservation, in the early 1920s, drew its support from myriad sources: women's clubs interested in questions of aesthetic beauty and moral uplift, sportsmen concerned with the propagation of fish and game stocks, and tourism promoters with visions of woodland riches, to name just a few. He was keenly aware that the well-meaning agitation of all these groups might succeed only in diverting attention from the hard economic realities of the cutover situation. So Lovejoy fashioned his argument as a no-nonsense appeal to the farmer, phrased in economic terms. Not once in his *Country Gentleman* writings did Lovejoy mention the beauty of the forest or the power of a wooded landscape to refresh the soul of an urban dweller. Even the term "forestry" had a "sort of taint" about it, he believed, mixed as it was with "old and sticky sentiment and sickly old-maid ideas." Blaming lumbermen for the waste of the North Woods served no good purpose, since it involved "moral considerations rather than economic."[29]

Lovejoy's mission was to lay out the cutover problem in clear, simple terms for a mass audience. By the end of 1922, he had largely succeeded. What remained was to link the cutover question to a larger crisis in land use in the United States, and to issue an unambiguous call to action. In "Settling the East," a seven-part series in the *Country Gentleman*, Lovejoy revealed that the Great Lakes states' problems with idle land were hardly unique: despite a barrage of promotion by state governments and private interests, some 200 million acres east of the Mississippi River lay vacant and unproductive. Aggregate data told the story of this swelling acreage in stark terms that boosterism could not refute. Just weeks after the series concluded, Lovejoy traveled to Menominee, Michigan, where he was keynote speaker at the Tri-State Development Congress for 1923. "The old ways and formulae and slogans have failed abominably," he told an audience of several hundred boosters, agriculturalists, and government officials. In Michigan, 600,000 acres had already reverted for nonpayment of taxes. Forest and farm development should proceed in tandem — but for anyone to hope that farms could possibly fill forty million acres in the Great Lakes states in the foreseeable future was a fantasy. Lovejoy's frank talk "proved a revelation to many," a reporter noted.[30]

Without gush or sentiment, he had issued a challenge to many of the prevailing ideas about cutover land use. To be sure, the old ideas — particularly the local emphasis on "town building" — would not yield easily. With the struggle in progress, Lovejoy could not be sure to what extent his ideas were being heard or accepted. In 1922 he summed up his experience for a colleague:

> It's a queer experience — this being out of the roil of forest things, looking in, trying to pick out the typical phases and make pictures of them; trying to discuss them without rancor and still accent whatever seems to need accent. Nobody can hope to do a perfectly balanced job of it, I suppose, and I shall be blamed and cursed and thanked tearfully no doubt — but don't much care — so long as the idea of timber-the-crop begins to get over, and especially to the farmers of the country.[31]

During his years as a journalist, Lovejoy had never really left the academic or professional realms. The cross-pollination of scholarly, professional, and popular ideas, in fact, had lent his writing a certain energy and insight that few other magazine scribes could muster. In 1923, professional conservation work called again: Lovejoy was asked to help organize the Michigan Land Economic Survey, a project closely mirroring his desire for data collection in the cutover. It was too good an opportunity to pass up, so Lovejoy left the tenuous enterprise of full-time freelancing in exchange for a brand of government work that was close to his heart.[32]

The Vision Fulfilled

"If words would make trees grow," the *Journal of Forestry* observed in 1926, "the United States would be the most thickly wooded country in the world." Everyone, it seemed, was talking of forestry by the mid-1920s — "Women's Clubs, Rotarians, Kiwanians, hundreds of Civic Clubs of every description, . . . sportsmen, nature lovers," even real-estate promoters and, lately, lumbermen themselves. Yet the destruction of the woods proceeded apace, reforestation was in its infancy, and efforts at systematic forest-fire suppression had proved barely adequate. The forestry movement, it seemed, had degenerated into a Babel of well-meaning talk without depth, coherence, or effectiveness. "We are off either on our economics or . . . psychology," the *Journal* concluded.[33]

Actually, what forestry needed was a convergence of harsh economic reality and diffuse public sentiment. Lovejoy had predicted as much in 1922, in a letter to his editor at the *Country Gentleman.* Americans had to be "shocked out of the old notion" that all vacant lands would yield to the plow and that individual initiative was sufficient to solve the land-use problem.[34] Remaking the cutover would require government-coordinated social engineering on a scale never before seen, except perhaps in wartime.[35] Public sentiment, though useful to the foresters' cause, focused mostly on individual acts. Voluntary, small-scale, feel-good efforts — such as Arbor Day tree plantings — would not suffice. Citizens would need hard evidence before assenting to an expansion of government authority and centralized planning in a landscape that had epitomized rugged individualism. Regrettably, the catalyst for meaningful action would have to be a crisis.

The crisis was tax reversion. In Wisconsin, whose experience was typical of the Great Lakes states, the population of the cutover actually was less in 1930 than it had been in 1920. Land values in fourteen of the eighteen cutover counties fell during the decade. Lagging crop prices meant that many of the settlers who had taken on debt during the land-selling boom after the war now could not make payments. The federal clamp-down on immigration had dried up a stream of potential pioneers. At the urging of land economists, the agricultural college and state government had stopped promoting settlement in the mid-1920s. In 1927, a mortgage banker reported that farm land in northern Wisconsin was "not saleable." Property tax bills, small as they were, became too much for landowners to bear. Slowly at first, then at an accelerating rate, the owners stopped paying their taxes. The lands reverted to county governments in Wisconsin and to the state governments of Michigan and Minnesota.[36]

By any reckoning, the magnitude of tax reversion was nothing less than staggering. Two and a half million acres of Wisconsin cutover land — about one-quarter of the land in seventeen northern counties — were put up for sale for nonpayment of taxes in 1927 alone. Only 18 percent of the land was resold. The rest remained with the counties, where it sat idle, fed forest fires, and produced not a penny in tax revenue. Large areas of the cutover seemed to have "little present market value for any purpose," a team of researchers reported. The growing tax delinquency meant that the cost of government services, such as schools and roads, increasingly was borne by fewer citizens. As a result, delinquency itself threatened to cause more delinquency, in a ruinous spiral whose end could only be guessed at. The land economist Benjamin Hibbard saw an ironic turnabout in the tax-reversion crisis: after years of giving away the public domain, governments now were getting that domain back, whether they wanted it or not. The challenge was what to do with it.[37]

For the most part, the tax crisis was caused not by farmers but by land speculators, who dumped their holdings when the futility of their endeavor became clear. But the avalanche of tax-reverted acres fueled the perception that cutover agriculture had run up against its limits. Some cutover farmers had survived by adapting to the peculiar circumstances of their environment — hunting and fishing for subsistence, bartering with neighbors, or working in town, relying on the market but refusing to be dominated by it. Such practices were hardly new. But in the midst of an agricultural depression, such "backward" ways convinced many observers that the farmers' persistence was illogical, misguided, even dangerous to society.[38] Lovejoy, who recognized the intractability of the tax problem earlier than most, wrote to a fellow forester in 1922 that without expert stewardship, the cutover was certain to go on "breeding paupers and morons and fires."[39] His attitude toward cutover settlers, like that of most of his colleagues, eventually would harden from paternalistic concern to deep dismay, even revulsion.

By the end of the 1920s, the nineteenth-century "land ethos" had unraveled in the North. Widespread land dispersal had not been an engine for economic growth or for the attraction of capital or labor. The promise of "town building" and its attendant capital gains in land had proved to be a cruel joke for the area's pioneers. In many places, the frontier of settlement actually was receding; thus the settlers who had hoped for an influx of social amenities increasingly found themselves isolated from neighbors and markets. And despite the push for incremental agriculture, the cutover had not become a land of farms.

Especially with the coming of the Depression, the Great Lakes states embraced reforestation and a host of related land-use innovations. These measures included public acquisition of northern lands for state and national forests; rural zoning to discourage scattered settlement in forest areas; and special tax abatements and state assistance to encourage the growing of timber as a crop by local governments and private owners. Government-financed fire control assured long-term protection for such previously overlooked assets as young forests, recreational ventures, scenic beauty, and wildlife. The value calculus of forest land use had changed dramatically, with government initiatives now augmenting private decision making to make the region self-supporting and socially stable. Rural zoning, for example, recognized the social overhead costs of providing roads and schools to widely dispersed residents in remote areas; it curtailed the individual's freedom to live where he wished in favor of a larger societal interest in economic efficiency. And governments by the late 1920s had come to recognize the forest's potential role in regional rehabilitation and economic stability.[40]

What role had Lovejoy played in this reordering of ideas? Even before 1920, Lovejoy had recognized that most conservationists were too isolated and too parochial, focusing on narrow matters of technical proficiency. Silviculturalists, for example, specialized in the growing of trees, but many of them lacked the political or economic knowledge to make winning arguments for wide-scale reforestation. Technical proficiency had to be joined with economic rationale and political muscle, Lovejoy believed, if the long-standing land ethos was to be toppled in the North. The ethos, in fact, might have lingered for years if not for the coming of an economic triple blow: the nationwide farm slump in the 1920s, the tax-reversion crisis in the cutover, and the onset of the Great Depression, with its attendant mandate for widespread planning and public-employment programs. A handful of conservation writers could not bring on reforestation by themselves. But they could tutor the public and the experts alike on what to do when the crisis became intolerable.

After 1923, Lovejoy never returned to full-time magazine work. Instead, he practiced his "banana-peel engineering" from within, as a staff member at the Michigan Department of Conservation. After working on the Land Economic Survey, he helped establish a system of game refuges. In 1930 he suffered a disabling stroke, and afterward his health was precarious. Until his death in 1942, Lovejoy served as a gadfly-at-large within the Conservation Department. His focus ranged from fisheries to forests to fire protection and public education. His friend and colleague Harold Titus summed up Lovejoy's contribution thus: "It was P.S. who was everlastingly leading and prodding until the plan was more than wish or lines on paper and had become reality."[41]

Still, his effectiveness might have been greater had he been willing to acknowledge, and exploit, the sentimental and mythical side of the conservation impulse. In retrospect, Lovejoy's aversion to sentiment seems excessive — he wrote that nature lovers and women's clubs "only queered us with the hard-boiled gentry about the statehouse" and that he preferred "a few well-informed Chambers of Commerce" to sportsmen's groups any day.[42] Zealous in his economic mission, he failed to see components of the cutover equation (including, possibly, a role for yeoman farming) that could not be expressed with numbers. His vast and insightful correspondence, outcroppings of which can be found in the papers of almost every major conservationist of his era, is oddly barren of any mention of Lovejoy's own feelings for the forest, which presumably ran deep. The reader is left to wonder whether a man so sure of his own intellect might have mistrusted his intuition.

In any case, Lovejoy's dismissal of sentiment was his major blind spot and the one factor that prevented him from transcending economics to enter the higher realm of environmental ethics. Always willing to tell his readers what to do with their environment in a given situation, he none-theless failed to divine any long-standing philosophy of humankind's relationship to nature.

For all that, his vision and his work are still impressive. "The parentage of ideas about men and land is seldom recorded at all," Aldo Leopold wrote shortly after Lovejoy's death. Leopold noted that any person could go to the Patent Office and discover the intellectual lineage of such mundane inventions as "egg-openers, iceboxes, and cigarette lighters." But the history of ecological ideas was not nearly as discernible. Still, Leopold offered a sweeping pronouncement on the subject: "I believe that P. S. Lovejoy sired more ideas about men and land than any contemporary in the conservation field."[43]

Two decades earlier, Lovejoy had entrusted the general populace with a set of profoundly radical ideas about the forest, confident that the public, if confronted with facts, would yield certain liberties to the crusade for organization and expertise. His driving principle, and his enduring contribution, was his insistence that conservation was not just a technical endeavor, but a social one as well.

CHAPTER FOUR

Cowboys and Bureaucrats

The Cult of the Forester

> Freedom! That is the key to the magnetic lure of forestry.
> Independence, danger, the brooding mystery of the forest, the strife
> of storm.... These different elements all appeal in varying degrees to
> your true forester, thus welding a solid bond between members of
> the clan.
>
> —CARLTON GORDON MURRAY, forestry student, 1928

By the time Carlton Murray graduated from Michigan State College in 1929, he would be what was then referred to as a "trained forester." It was still a relatively small field. Besides Murray, 290 others in the United States would receive bachelor's degrees in forestry that year.[1] During his years in East Lansing, he was exposed to several basic disciplines: silviculture (the growing of trees), entomology (the study of insects), biology, botany, chemistry, soil science, and agricultural economics. Summer jobs in the national forests acquainted him with forestry's everyday work: timber cruising,[2] reforestation, handling of timber sales to private cutters, fire fighting, and dealing with a sometimes skeptical public.

It was an exciting yet tense era for forestry. Upon graduation, Carl Murray probably sought permanent employment with the U.S. Forest Service (USFS), which was busy acquiring cutover land for new national forests east of the Mississippi River. In addition to their routine work, foresters were grappling with their profession's growing pains. There were conflicts over the use of forest resources. Did they exist for timber or for tourism? There were chronic complaints over the low pay and poor housing for USFS field staff.[3] When a "college man" such as Murray advanced from the field into an administrative job—and most of them did so quite quickly—the pay rose above bare subsistence but the pressures multiplied. So great were the stresses of administration that the Forest Service chief, Robert Y. Stuart, suffered a nervous breakdown in 1932. In 1933, Stuart plunged seven stories from an open window at his Washington headquarters, an apparent suicide.[4] Like many other jobs, forestry could destroy people who were not adept at navigating a modern, hierarchical organization. Knowing one's trees was not enough.

Yet Murray, in a prize-winning college essay reprinted in the *Journal of Forestry*, painted the forester's life in roseate hues. Montana, where he had worked as a student forester, was a "wilderness" with links to the Old West. Every true forester was a "dreamer of the wilds." The fabric of a forester's existence was woven of "jolly nights" in French-Canadian villages or manly camaraderie around the campfire. The forester's mission invoked the lore of lumberjacks, or of the Canadian North, "free, savage, crude—the North where the strange sky-beacons glow over the loneliness of a frozen land." Murray made clear that the stresses he encountered would not be bureaucratic, but would emanate from the age-old clash between solitary man and raw nature. "Life is not placid, not easy. Give me the scene of actuality, storm!"[5]

On its own, Murray's epistle might be dismissed as youthful naïveté. But during the 1920s, his was just one voice among many to celebrate the romantic life of the forester. Most of these authors were older than Murray. Many were practicing foresters with experience going back decades. Some of the most prominent spinners of forest-ranger tales were high-ranking Forest Service officials. Will C. Barnes, for example, wrote forest nonfiction for top magazines such as the *Saturday Evening Post*. His day job was conducted not in a saddle but behind a desk, as manager of grazing for the USFS in the West. Yet Barnes peopled his stories with independent, solitary men, not unlike the heroes of cowboy tales.[6]

The obsession for stories of the forest—in short fiction and nonfiction for adults, in juvenile serials, in novels, on film—began in earnest about 1920. Earlier writers had enjoyed tremendous success with out-of-doors books. Ernest Thompson Seton, for example, had climbed the best-seller lists with his fanciful (and much criticized) stories of animal life, such as *Wild Animals I Have Known* (1898).[7] But the early works seldom dealt with the day-to-day life of forest regions. As late as 1914, a magazine editor had suggested to the Michigan writer Harold Titus that he might try finding material in his own back yard, in the form of "good stories of the Michigan lumbering camps."[8] Titus — like other writers — eventually caught on. So did readers. The novels of James Oliver Curwood, set in the wildlands of Canada, would explode in popularity about 1919. A craze for stories of "the great open spaces" made Curwood the second most popular novelist of the period 1917–26.[9] The literary pack was led by Zane Grey, a former Ohio dentist whose cowboy novels, in the words of Roderick Nash, "emphasized courage, self-reliance, fair play, and persistence — the traditional frontier virtues."[10] Forest tales emphasized such traits as well, especially when translated into silent film. Apparently it was almost impossible to go to the movies during the 1920s without seeing a picture in which a dashing young woodsman rescued a girl from a forest fire.[11]

But beyond the whims of popular culture, what did it mean? It is highly significant that the authors who celebrated the forest were often foresters themselves. Writing about themselves — indeed, crafting their own myth — these authors sought to explain, and glamorize, a vocation that bridged tradition and modernity. The forester acted in a "natural" environment, yet he was on the forefront of organization in the use of natural resources. He relied on his wits, but also on science. In an age when cowboys were relegated to sideshows, his occupation represented a chance for men to test their mettle and assert their leadership in the open spaces. Foresters were the advance guard of a new bureaucracy that would administer the woodlands in the name of politically agreed-upon social goals. They were individual agents of collective, centralized action. Like the heroes of the cowboy tales, they acted alone but portended the coming of the forces of law. Still, because of their affinities and their working environment, it was relatively easy to portray them as individualists. The foresters of 1920s popular culture were agents of modernity riding horseback. Here as elsewhere, the mythology of individualism would be used to garner public support for activities that were essentially collective in nature.[12]

When the "Old Guard" Was New

The American forester was a creation of the early twentieth century. His job involved equal parts law enforcement, conservation, administration, and public relations. The burden of bringing law to the nation's earliest parks and forest reserves had fallen to the military and to a small group of cowboy-rangers working for the Department of the Interior. But in 1897, Congress had specified that the forest reserves of the West should be managed, not just protected against trespass or timber theft. Foresters suddenly had a mandate for broader action.[13] The nation's first forestry school was established at Cornell University in 1898, and a master's program at Yale University — endowed by Gifford Pinchot's family — soon followed. Pinchot became chief of the Agriculture Department's Division of Forestry in 1898. One of his earliest moves was to establish a student intern program, bringing well-born young men into the office for training before sending them into the woods. As one of Pinchot's associates noted, there was some need to "get the Harvard rubbed off the students before they came in contact with the loggers." The intern program was the nexus of an intense esprit de corps among forestry's "Old Guard," as the students came to call themselves. The Ivy Leaguers' enthusiasm for outdoor life was so great that three hundred students had entered the program by 1902.[14]

Under Theodore Roosevelt, Pinchot consolidated most federal forestry activities in the new U.S. Forest Service in 1905. That same year, the Forest Service issued its first *Use Book,* so named because the forests were intended to be used for human benefit. The *Use Book,* which came to be celebrated as the ranger's bible, codified Forest Service beliefs and policies. It was small — just over four by six inches — so it could be slipped into a pocket or saddlebag. Properly managed forests, it stated, provided not only timber but also protection against encroachment of monopoly power. "They are patrolled and protected, at Government expense, for the benefit of the Community and home builder."[15] The *Use Book* represented Pinchot's determination to standardize and rationalize Forest Service activities under the ethos of "wise use" of resources. He allowed his subordinates considerable latitude in day-to-day management; given their isolation in the field, he often had no choice. But in matters of policy and discipline, the lines of authority went directly to Washington. As created by Pinchot, the USFS forester was "a new breed of government servant," in the words of one historian, a far-flung agent whose conduct was regulated by clearly articulated policies and an almost cultlike sense of professional fraternity.[16]

Although the college men of the "Old Guard" appear to have sincerely craved a red-blooded experience, differences of rank and background did not go unnoticed in the West. The rangers themselves tended to be recruited locally in forest communities, with a premium being placed on practical experience rather than "book learning." They had to pass a customized civil service exam that included shooting and using an ax. College-educated foresters usually began as forest assistants, learning administration in tandem with nuts-and-bolts fieldwork. They, too, had to pass exams and to prove that they were up to the physical rigors of the job. Yale men were rumored to get favorable treatment, but Pinchot worked assiduously to make the administrative ranks a meritocracy — albeit one of relative privilege.[17] Once in the field, the new foresters donned chaps and spurs and patrolled their beats on horseback. Good relations with the locals were essential. In a light verse, one forest assistant warned his fellow pipe-puffing "collegians" to soft-pedal their credentials and to avoid condescension in the woods:

> Get along with the men that you find on the job;
> Don't criticize grammar, and set up for a snob;
> They were woodsmen ere you learned to puff at a "cob"
> And wear a badge like a forest assistant.[18]

By all accounts, the forest assistants treasured their rough-riding days, clinging fondly to the memories long after they had advanced into administration or academic work. Among them were forester-editor Ovid Butler (Yale, M.F., 1907); game manager, writer, and ethicist Aldo Leopold (Yale, M.F., 1909); and forester-writer P. S. Lovejoy, who attended the Forestry School at Michigan before joining the Forest Service in 1905. Tellingly, all three of these men would spend the bulk of their professional lives not in the forests, but in centers of political and academic influence: Butler in Washington, Leopold in Madison, Wisconsin, and Lovejoy in Ann Arbor, Michigan. Especially in the 1920s, a growing concentration on research and legal-economic measures meant that the work of remaking the forests would be done in the cities. But in championing new ways, the members of the "Old Guard" would not forget their happy heyday in the woods.

Portrait of a Ranger

To say that the forest literature of the 1920s was "realistic" is not to say it was completely true to life. Writers in this era employed a sort of selective realism, emphasizing aspects of forest life that held a certain romantic appeal.[19] The imperatives of organization held no charm for the average reader; nor did the intricacies of land economics or timber taxation. As Ovid Butler had discovered, even talk of a "timber famine" was not sufficiently enticing to arouse the public imagination. Instead, the idea of forestry would have to be sold largely through pioneer and cowboy ideals: individualism, flinty self-reliance, bravery, and cheerful resourcefulness. All of these characteristics were embodied — or rather, came to be embodied, through conscious mythmaking — in the person of the forest ranger. (Popular writers, blurring the class distinction between rangers and college-educated foresters, tended to lump them together.) A sampling of the ranger literature reveals a number of recurring themes.

The Ranger Is a Cowboy

All of the early forest reserves were located in the West, often in rough terrain that had not been extensively settled. Much of the ranger's day was spent negotiating with ranchers over grazing rights in the forest. Because the early rangers were recruited from the ranks of ranch hands, it was easy enough to see the ranger as just another sort of cowboy. With the frontier gone, the ranger could be invested with all the mythical qualities of the gunslinger: independence, a roving disposition, and a certain scorn for civilized ways. In 1911, one unusually prescient writer forecast the coming tide of ranger tales in popular culture. With the Wild West tamed, the adventuresome soul had "one chance — one hope" left to him, and that was the forest. It was only a matter of time, the writer prophesied, before some genius would produce "his rendition of the ranger in some thrilling new play — some melodrama of the West."[20] Under this conception, the forests themselves became a sort of untamed frontier where cowboy ways could find full expression. "They say the Old West is gone — except in the movies," one magazine scribe rhapsodized in 1925. "But they forget the great stretches of mountains that progress has skipped."[21]

Ovid Butler, who earned his spurs in the Boise National Forest in Idaho, never forgot his own cowboy days. Editing the American Forestry Association's magazine two decades later, Butler ran a constant stream of western ranger tales, both nonfiction and fiction. The true stories eventually were collected in a book, *Rangers of the Shield*.[22] Butler commissioned a fellow forester, Wallace Hutchinson, to write stories for young readers featuring "Ranger Bill." The fictional hero wore a Stetson, rolled his own cigarettes, rode a horse named "Buck," employed improper but colorful grammar, and referred to forest fires as "ornery critters." Unlike Ranger Bill, Hutchinson was a Yale man (M.F., 1903) and chief of public relations for the U.S. Forest Service in California.[23] Besides his appearances in print, the Ranger Bill character later would become a regular on network radio.[24]

The Ranger Acts Alone

Pictures accompanying magazine articles of the 1920s frequently showed the forest ranger on horseback, literally riding off into the sunset. Writers made occasional mention of the Forest Service's "efficiency" or "businesslike" organization, but they seldom dwelled on it. More typical was an article in Ovid Butler's *American Forests and Forest Life,* celebrating "the solitary ranger, whose duty sends him into the back country to combat single-handed all the forces of nature."[25]

Even in fire fighting, an endeavor where cooperation and organization were essential, writers tended to glorify the battle between solitary ranger and savage blaze. A ranger might have to "cope single-handed with a fire that may be burning over an area the size of New York or Chicago," one account noted.[26] Magazine writers were especially fascinated with forest-fire lookouts, those lonesome souls who sat on mountaintops day and night watching for distant smoke or flames. One reporter offered a glowing profile of "Miss Lorraine Lindsley" of Wyoming, one of only two female fire lookouts in the Forest Service. Living alone in a tiny log cabin, she rode a white horse to her station each morning, reaching "the top of the mountain while many city dwellers are still at their coffee."[27]

The Ranger Faces Danger Frequently and Dutifully

Profiling William R. Kreutzer, supervisor of the million-acre Colorado National Forest, a writer seemed to marvel that the intrepid ranger was still alive. Born in a log cabin, Kreutzer possessed "a quick finality in speaking, developed through years of dealing with personal danger." Working alone, he had outfoxed gangs of gun-toting desperadoes in the forest. "It is an extraordinary thing that, with all the close scrapes he has had, Kreutzer has never been shot at, and has never shot at anyone." Even while speaking of near-death, the ranger related the details of his adventures "cheerfully," the writer noted.[28] Forest fires, in particular, gave rangers an opportunity to face danger squarely and to exert leadership over lesser men. Kreutzer had run through walls of flame to rescue his subordinates. In an incident that became legend, E. C. Pulaski saved a group of firefighters from a raging inferno in Idaho by herding them into an abandoned mine shaft. When the "panic-stricken" men tried to flee, he drew a revolver and threatened to shoot them. Without Pulaski's leadership the entire group would have been incinerated. The ordeal left him blind for two months.[29]

The Ranger Is Educated—but in a "Practical" Way

"To be a successful ranger, a man must have a working knowledge of many trades, from stringing telephone wires to building bridges," one article reported in 1925. The accompanying photographs made rangers appear to be overgrown Boy Scouts, tackling merit-badge skills such as flashing Morse code messages by "sun telegraph."[30] Many writers implied that forestry was a seat-of-the-pants affair that could be learned mostly in the field. These journalists tended to celebrate the "old-time" foresters, those native westerners who had joined the ranks before 1900. The aforementioned William Kreutzer had been a ranger since 1898. He told his admiring interviewer how he had taken correspondence courses in law, then had heard of the new forestry program at Yale. "Going to Yale was out of the question; but I got a friend to give me a list of the books used in the course, and I bought them, and read and reread them all, chiefly at night by camp fire and candle light."[31]

In such ways was the reader of the 1920s introduced—gently—to the world of modern forestry. New ideas were phrased in terms of old and comfortable myths. And they were delivered by familiar and sometimes crusty figures. Reforestation, game protection, or a land-economic survey would be more palatable to the public when explained around a campfire. In this respect, it is not surprising that writers often employed the forest ranger as a sort of rough-hewn professor, delivering folksy lectures on conservation themes.

Slowly, the most astute writers were realizing that Americans could be taught to cherish the forest largely for the myths it invoked in the public mind. The forest—much like the Old West—could become a mythical landscape, a repository for the nation's image of itself as shaped through mass media. If Americans viewed the forest as a reflection of their own identity, they would be more willing to protect and propagate it. And what better embodiment of American ideals, what better traditional ambassador of modernity, than the hardy ranger? As one writer put it, he was "the link which unites a glorious past to a practical present."[32]

Confronting the Frontier

Historian Richard Slotkin has written that popular thinking about the frontier, circa 1900, tended to fall into two camps. One, which Slotkin labels the "Populists," mourned the passing of the untamed West and fretted about the future of a society without the hardening quality of western experience. The most famous frontier historian, Frederick Jackson Turner, belonged to this group. The other camp, which Slotkin calls the "Progressives," saw the frontier as a proving ground where the rightful American elite had risen to leadership. The frontier, in this conception, had allowed the emergence of a natural aristocracy. But the western frontier was not to be mourned, because many new frontiers lay ahead: in commerce, in politics, in worldwide adventurism and war. The most famous exponent of this view was Theodore Roosevelt. TR shot grizzlies, rode with cowboys, and slept outdoors in snowstorms, but his wilderness encounters were essentially a metaphor for a brand of joyful vigor that could just as easily be applied in public life.[33]

Popular novels of the forest mirror Slotkin's duality quite clearly. Turn-of-the-century tomes either mourn the forest's passing (without quite knowing what to do about it) or celebrate the lumber camps as a place where natural leaders can emerge to conquer the wilderness. More than a collection of trees, the fictional forest becomes a mythic space: it is a repository for Americans' ideals about their past and anxieties about their future. The "Populist" and "Progressive" views of the forest foreshadow the tension between individualism and organization that would characterize the remaking of the cutover in the 1920s.

What is most interesting, in this sense, is how some writers of fiction would eventually reconcile this tension in their work. In the 1920s, a handful of popular writers produced a distinctive subgenre of cutover polemical novels, emphasizing the need for scientific forestry. These books, which tend to be didactic melodramas, speak highly of expertise and organization. But the forester-heroes of these works are thoroughly individualistic. They act alone. They are independent of bureaucracy, usually through some happy accident of fate which has left them wealthy. Unlike most real-life experts, they actually live in the forest communities that they are striving to remake. They are educated (or closely allied with someone who is), but they downplay "book learning" and emphasize the "practical" aspects of forestry. When called upon to prove themselves, they do so in individualistic ways—such as winning fistfights with lumberjacks. And in remaking the forest, they always portray it as a place where the individual can find moral regeneration and fulfillment.

Does this somewhat awkward reconciliation—forged by authors who unabashedly championed scientific forestry—constitute a "hegemonic" crusade clothed as popular fiction? Yes and no. It could be argued that cutover novelists of the 1920s were only aping the language of societal elites who sought to make the forest continuously productive. Most cutover writers adhered to what Samuel P. Hays has called the "gospel of efficiency": the idea that conservation was an exemplar of a larger program of social management, a campaign to "bring conscious foresight and intelligence into the direction of all human affairs."[34] But to deride the cutover polemicists' writings as merely disingenuous obscures the cross-currents of motivation and sentiment that pervaded the cutover debate, including conflicting sentiments held by the writers themselves. A blunt-force methodology on this topic can only obscure the fact that cutover writers were grappling, quite seriously, with defining ways in which state action might facilitate a modicum of individual enjoyment in the woods.[35]

Four popular novels will illustrate the changing contours of cutover literature as the century progressed. In them, one can see the opposing poles of frontier thought as applied to the forest at the turn of the century, as well as the accommodation between individual and organization that marked the 1920s.

The "Populist" Frontier in the Forest: Silas Strong, 1906

Irving Bacheller was the Sunday editor of Pulitzer's *New York World*. He left newspaper work in 1900 after publishing *Eben Holden*, a phenomenally successful novel about pioneer life in upstate New York. Building on his fame, Bacheller produced a steady stream of fiction about New York's Adirondack and Saint Lawrence River regions. His frontier novels displayed a "cheerful sense of life," one critic wrote, with an emphasis on historical allusions and moral goodness.[36]

Bacheller's essentially romantic conceptualization of the woods did not fully equip him to deal with forest destruction. In 1906 he published *Silas Strong, Emperor of the Woods*. To the modern reader, *Silas Strong* is a rambling and rather unfocused tale of an Adirondacks fishing guide and his eclectic acquaintances. The namesake of the novel is a stammering primitive who seldom speaks more than two words at a time. Strong apparently is supposed to possess some sort of higher moral wisdom, though the nature of that wisdom is not made particularly clear. In one rare moment of lucidity, he scolds a timberman who is clear-cutting the forest: " 'N-no, no,' he said, 'it can't be. Ye ain't no r-right t' do it, fer ye can't never put the w-woods back agin. . . . God planted these w-woods an' stocked 'em. . . . Y-you ain't no right t' git together down there in Albany and make laws ag'in' the will o' God.' "[37]

Caught in a losing battle with the timber interests, Strong is killed at the book's end. His beloved forest seems fated for destruction. Bacheller tips his hat to Progressive Era conservation by including an articulate politician, Robert Master, among Strong's allies. But Master, too, conceives of the forest in terms of good and evil, with "Satan" standing behind the throne of "Business."[38] The book places scant faith in human solutions to the forest problem. The author's ambivalence about human entanglements is best seen in the beautiful and mysterious Edith Dunmore, who is held captive in the woods by her father for fear that civilization will corrupt her. Emphasizing the book's portrayal of the woods as a seat of pantheistic mysticism, Bacheller refers to Edith as a nun whose cloister is the forest.[39] She and Robert Master eventually marry, but Bacheller introduces the child-woman to the larger world with great reluctance. As for Bacheller's monosyllabic hero, he will get his reward only in heaven — a place where, it is presumed, the forests are green and lush. "He was never to bow his head before the dreaded tyrant of this world."[40]

The "Progressive" Frontier in the Forest: The Blazed Trail, 1902

Stewart Edward White was the son of a millionaire lumberman from Grand Rapids, Michigan. He earned bachelor's and master's degrees from the University of Michigan and attended Columbia Law School. In his twenties, he worked as a cowboy and shot wild game for mining camps in the West. In 1901, while living in a lumber camp in the Upper Peninsula of Michigan, he wrote *The Blazed Trail*. Published in 1902, the novel went through fifteen printings by the end of 1904.[41]

The Blazed Trail tells the story of Harry Thorpe, an educated and ambitious young man who clashes with the forces of raw nature and human greed amid the pines. It is set in the 1870s and 1880s. At a lumber camp in Michigan's Lower Peninsula, Thorpe quickly emerges as a natural leader of men. He moves to the Upper Peninsula, where vast stands of virgin pine await the coming of civilization. After building an enormous business, he is destroyed in a battle with unscrupulous timber-grabbers, but maintains his dignity. He has won the respect of his men, which is more important than money. As the novel ends, he is determined to fight anew. The book scarcely mentions conservation.[42]

White's key theme is leadership. In the forest, the right to lead is not bestowed through mere pedigree, but must be earned. "The lumber-jack will work sixteen, eighteen hours a day... for the right man! Only it must be a strong man — with the strength of the wilderness in his eye."[43] Even the trees have a pecking order — the pines are "vast, solemn, grand, with the patrician aloofness of the truly great."[44] Success is not an end in itself, but an ongoing duty. The "love of ease" is a sin. The getting of money is pointless, and sordid, unless it is allied with "some great and poetic excuse," such as clearing the way for a "higher civilization." "That is the only sort of aristocracy, in the popular sense of the word, which is real; the only scorn of money which can be respected."[45]

As a student in the 1890s, White almost surely had been exposed to Turner's frontier thesis of 1893. In a vignette at the beginning of *The Blazed Trail*, he pays tribute to the pioneer in a manner that starkly mirrors Turner's argument. The frontiersman is "[r]esourceful, self-reliant, bold; adapting himself with fluidity to diverse circumstances.... Stripped of all the towns can give him, he merely resorts to facile substitution. It becomes an affair of rawhide for leather, buckskin for cloth, venison for canned tomatoes.... In him we perceive dimly his environment."[46] Yet unlike Turner, White is scarcely moved by the passing of the frontier. Instead, he sees the pioneer's raw passions (including his "wild excesses")[47] as a picturesque but transitory element in a landscape that is evolving from savagery to civilization. What matters is leadership — specifically, the emergence of a natural aristocracy in an environment of strife and struggle. The forest would be cleared soon enough, but the quest for the allegiance of men could be carried out in new settings, on new frontiers. Harry Thorpe is not a woodsman, but a soldier:

> With him the great struggle to wrest from an impassive and aloof nature what she has so long held securely as her own, took on the proportions of a battle. The distant forest was the front. To it went the new bands of fighters. From it came . . . men groaning on their litters from the twisting and crushing and breaking inflicted on them by the calm, ruthless enemy. . . . Here at headquarters sat the general, map in hand, issuing his orders, directing his forces.[48]

Not surprisingly, the young writer caught the attention of President Theodore Roosevelt, who devoured a copy of *The Blazed Trail* in a few sittings and was "delighted with it." White visited TR soon afterward, and the two men cemented their spiritual kinship on the White House tennis court.[49] In 1904, Roosevelt deputized White as special inspector for the California Forest Reserves. The president also lauded the results of a hunting trip in which the novelist had slain 105 wild boar, some of them with a knife in close combat.[50] A skilled marksman, White was a favored guest at shooting matches at the Roosevelt estate in Oyster Bay — where, tellingly, most of the competitors were not frontiersmen, but military leaders.[51]

The Individual and the Organization: Connie Morgan in the Lumber Camps, *1919, and* Connie Morgan with the Forest Rangers, *1925*

James Beardsley Hendryx was born in Sauk Centre, Minnesota, in 1880, five years ahead of the village's more famous literary son, Sinclair Lewis. After two years at the University of Minnesota, Hendryx worked as a traveling salesman, cowboy, and Cincinnati newspaperman. Beginning in 1915, he turned out a number of adventure novels set in the North Country of Alaska, Canada, and the Great Lakes states. These books included a juvenile serial featuring a plucky boy hero named Connie Morgan.[52]

Connie is a character straight out of Horatio Alger: an orphan who starts out "forlorn and friendless" in Alaska and makes his own way in the world. Fortunately for him, the implausible serendipities of popular fiction help smooth his path. He meets up with a kindly father figure, "Waseche Bill," and makes a quick fortune in the gold fields. By the time he is twenty, he possesses a worldly resourcefulness, a keen sense of the measure of men, and enough money to do as he pleases. He rides with

Mounties in Canada, matches wits with timber barons, then takes on the organizational challenges of modern conservation work.

In *Connie Morgan in the Lumber Camps,* Connie travels to Minnesota to check up on a timber tract he co-owns with his Alaska partners. In the forest, he must battle the dual enemies of labor agitation and corporate corruption by the German-owned timber "Syndicate." After single-handedly dispensing with the hooligans of the Industrial Workers of the World, Connie journeys to Minneapolis. There he outfoxes the imperious German officers of the "Syndicate." Having negotiated a clever contract, Connie returns to the woods to deliver his timber personally. The hero leads a heart-stopping log drive down a swollen river, during which his chief rival drowns.[53]

What is amazing—and hugely improbable—through all these travails is how Connie manages to straddle two worlds: the old world of pioneer individualism and the new world of capital, organization, and management. He is modest, resourceful, and unfailingly cheerful, and he prefers to act alone. To win the respect of the lumberjacks, he proves himself the old-fashioned way—with his fists. Yet he is clearly a natural leader, fit for the business world. He possesses a distinctive middle-class bearing, as opposed to the slouching and unshaven men of the lumber camps.[54] As he walks purposefully through the streets of Minneapolis, passersby notice "a certain something in the easy swing of his stride, the poise of his shoulders, the healthy bronzed skin and the clear blue eyes."[55] Setting foot in a corporate suite for the first time, he gets the best of the timber titans in a tense negotiating session. Despite this business acumen, Connie is almost illiterate: he is flummoxed by a restaurant menu and invites the waiter to join him for some grub. Always, he adheres to the code of Alaska, which requires that a man be judged by his deeds, not his rank. "Up where I live they don't call a man 'sir' just because he happens to have a little more [gold] dust than somebody else. It ain't the 'Misters' and the 'Sirs' that are the big men up there; it's the 'Bills' and the 'Jacks' and the 'Scotties.' "[56]

Published six years later, *Connie Morgan with the Forest Rangers* is attuned to the specific technical challenge of remaking the woods. Arriving in Pine Tree, Michigan, the fictional hero buys twenty-three thousand acres of timber and announces his plans for sustainable forestry and selective harvest. Here, Connie has all the support needed for administering large tracts of land. He reflects the power of the state (through a commission as a game warden) and the expertise of scientific forestry (through the presence of an older sidekick, a forester named McLaren). Still, Connie acts independently, with an unlikely combination of business skill and impetuous cleverness. Once a boomtown, Pine Tree has gone to rot, and the forest has been devastated by rampant greed and waste. Connie and McLaren represent the avant-garde of expert guidance, with McLaren delivering a series of soliloquies on the need for reforestation. Yet at each critical juncture, Connie does things his own way. His chief adversary is John W. Crump, a scheming timber rapist intent on clear-cutting the forest. Foiling Crump at every turn, Connie negotiates a series of complex land deals, buys a railroad, and finally dynamites a dam to flood the forest and save it from a fire that Crump has set.[57]

Much like the real-life effort to remake the cutover, Hendryx's tale reflects a tension between individualism and organization. Most of the citizens of Pine Tree are unschooled in reforestation; they shrug their shoulders when the land burns. A few pathetic farmers labor at the mercy of their mortgage-holders, who speak foolishly of an agricultural paradise to come. Pine Tree, with its ramshackle buildings and dirt streets, represents the tragedy of individualism run riot. The regression of the landscape is replicated among the people, who are slack-jawed, listless, and venal. "When the timber's gone the country ain't no good, and the folks that lives in it sort of gits no 'count, too," declares the local sheriff. ". . . I tell you the cut-over is an unmoral place to live in."[58] In this hellish and hopeless landscape, Connie and McLaren preach the gospel of organization and knowledge. ("That's the trouble—ignorance!" McLaren says at one point.)[59] But organization is always portrayed as a means by which the individual can flourish. In *Connie Morgan with the Forest Rangers,* this is seen through the character of Clayt Mimms, a corpulent swindler who is Pine Tree's corrupt justice of the peace. After going to work in Connie's progressive forest, Mimms emerges fifty pounds lighter, with a newfound gleam in his eye. The redeemed man announces that Connie has given him the "Cedar Cure."[60]

By the mid-1920s, the cultural groundwork for the cutover revival had been set in place. The reimagining of the forest would be a delicate balancing act, fraught with tension and occasional contradiction. Through the magic of expertise, coordination, and compulsion, the cutover would again become a place where (some) individual dreams could flourish. Organization would serve the ends of the independent spirit. The best cutover publicists certainly recognized the movement's conflicting strains of thought. Those who combined popular writing with conservation politics had to be especially self-aware. They tailored their rhetoric to the needs and capacities of their audiences, moving incrementally, building a complex edifice of science, law, methodology, and myth. Life was not a storybook, and vice versa. The writer who forgot that lesson might hang himself with his own words.

CHAPTER FIVE

James Oliver Curwood and the Limits of Antimodernism

Why should I not write of ideals — of the things we all aspire to be — instead of clogging what little ability I may have with the mud and slime of what we do not want to be? . . . The world is filled with men as strong and good, and women as beautiful and virtuous, as any of those who live in the pages of my books.

— JAMES OLIVER CURWOOD, *The Glory of Living*

This Curwood has spherical hypertrophy of the ego. One can touch him only on a tangent. . . . He operates on the emotions.

— P. S. LOVEJOY

In 1921, Hearst's Cosmopolitan Book Corporation published a confession titled *God's Country: The Trail to Happiness*. Written by a former hunter, it told how the man had renounced killing after finding God in the forest. It told how he had shed his "egoism" and his belief that he must dominate the Earth. It told how he had rejected the "superstition" of organized religion. The encounter with God, the author declared, was as simple as a walk in the woods. "I have found the heart of nature. I believe that its doors have opened to me, and that I have learned much of its language."[1]

The author was not a mystic or a seer. He was a novelist. His name was James Oliver Curwood. For more than a decade, Curwood had captivated readers with romantic tales set in the frozen North Country of Canada and Alaska. As hunter and adventurer, he had visited most of the places he had written about. Such was his popularity that, in the

James Oliver Curwood. Like his fictional hero Philip Steele, he was constantly looking back "to almost forgotten generations . . . when romance and adventure were not quite dead." Photograph courtesy of State Archives of Michigan.

1920s, he would be ranked as one of the highest-paid writers in the world.[2]

Now, with the revelation of his personal epiphany, he implied that fiction and fact had fused into a blissful union. In *God's Country*, Curwood cast himself as the hero of his own romantic journey. The book, he stated, had been written at a wilderness cabin 1,500 miles north of his Michigan home. As he typed out the words of his creed, the birds had laughed and swooped around him; a menagerie of other creatures had skittered into view as if to commune with their newfound ally. He was alone, and utterly content, in the garden of nature.

Yet in many ways, the book was a lie. Its wilderness authorship was a fabrication. Curwood had not made an extended trip to the Canadian backcountry in several years.[3] Riding a crest of fame, he had become a literary deal-maker and film producer. In the process, he had become ensnared in the machinery of business life, and the irony of this captivity gnawed at him. Though he did find occasional solace in the Michigan woods, he was a thoroughly conflicted man. His effusive correspondence reveals him to have been expansive and vain, given to sweeping enthusiasms and intemperate outbursts. His health had broken down as early as 1911; by 1920 he was continually on the verge of incapacity. "[I]t is our nerves that kill us in the long run, our over-restless minds, our worrying, questing brains," he wrote in *God's Country*, probably speaking more for himself than for the human condition.[4]

In 1921, Curwood entered a new arena of conflict, in which his romantic conception of his own life clashed head-on with reality. The arena was conservation. Crusading for wildlife, Curwood declared his opposition to all that was "political." Believing that the reasons and methods for conservation were self-evident, he damned as an enemy anyone who disagreed with him. After years as an outsider, he was elevated to the Michigan Conservation Commission in 1927. By then, his vanity had blinded him to the fact that well-meaning people could disagree. Six months into his tenure, Curwood — with eerie prescience — was casting himself as a martyr. Only his death, in August 1927 at the age of forty-nine, spared him from total ruin as a policy maker.

His demise also put a premature seal on his reputation. Lionized by many, Curwood has been remembered as an outdoor Renaissance man — equally adept as adventurer, author, filmmaker, and conservationist. He has never received an honest and unblinking historical treatment, from either popular writers or scholars.[5]

In the case of conservation, much can be revealed by such an inquiry. Curwood was a thoroughgoing antimodernist; as such he was poorly equipped to understand or advance the conservation methodologies of the 1920s.[6] At the same time, however, his nostalgic sentiments character-ized much of the tension inherent in remaking the landscape of Michi-gan and elsewhere. Conservation was a forward-looking movement of organization and method. But the very essence of the work—preserving and restoring an environment associated with the wild past—often dic-tated that conservation would be "sold" to the public through associa-tion with ideas that were distinctly backward-looking. In this sense, while Curwood failed as a conservation policy maker, he may have been con-servation's best "salesman" of the decade. As an author of fiction, he contributed enormously to the task of "reimagining" the landscape, but erred badly in thinking that he understood the process and all its contradictions.

His novels—dramatic, overwrought, and hugely popular in their day—resonate with the same antimodernist ideas that informed his conservation work. With Curwood, the literary conventions of romance and melodrama were more than a means to a paycheck. They constituted a worldview. Good and evil were plain to see, and the route to achieve-ment lay in self-mastery and individual heroism. The wilderness pro-vided an environment where valiant men and virtuous women could claim their destinies free of societal encumbrance. Curwood saw an un-breachable divide between human beings and nature, and was deeply ambivalent about the coming of "civilization" in any setting. Rejecting literary naturalism, he idealized the wilderness as a setting where small numbers of people might find happiness through their own energies.

Appealing as it was in fiction, this scheme was grossly unsuited to conservation work in the 1920s. Curwood's emergence as a world-famous author—and as a spokesman for the conservation movement—coin-cided almost perfectly with the years of greatest ferment in the remak-ing of the Great Lakes cutover. But while experts such as P. S. Lovejoy were divining the workings of an entirely new conservation apparatus, Curwood preferred to look to the past, seeing the forest as a stage for individual heroism and drama of the type that had been wiped out by modern living.

To be sure, nonspecialists could participate in the conservation movement, and often did. The most successful of these activists championed
collective measures by invoking the mythology of the frontier past. Curwood was not the only writer-activist to celebrate the scent of pines or
the lore of the voyageurs while trying to influence the policies of his
own day. But the path to success was strewn with hazards. Citizen activists — one historian calls them "radical amateurs" — had a long and
distinguished record of shaping conservation policy, largely through
their writings. But their moral fervor did not always translate to the policy sphere.[7]

Here, Curwood failed by insisting that truth in nature was divinely
revealed instead of humanly constructed. He stated that the "voice of
nature" spoke to him personally and would speak the same way to anyone who would listen.[8] He was oblivious to the theme of continuous
production and impatient with its details. Curwood never distinguished
myth from hard fact. His conservation crusade thus became a blunt instrument, without nuance or any awareness of irony or contradiction.
In real life, Curwood the storybook hero seemed fated for a tragic ending.

A Storyteller's Story

James Oliver Curwood was born in the small Michigan city of Owosso
in 1878. His father was a cobbler. The family moved to Ohio when Curwood was a toddler, but he returned to Owosso at age thirteen to live
with his older sister and attend school. His parents soon followed. Curwood was to spend most of his life in Owosso, and he often referred to
the city fondly. He left reluctantly to enter the University of Michigan,
having bluffed his way through the entrance exams without benefit of
a high school diploma. Evenings, he wrote adventure stories, a passion
he had pursued since childhood. Some of his yarns were published in
newspapers and in an obscure children's magazine, the *Gray Goose*. He
left college his junior year to work as a reporter in Detroit. His first novels, *The Courage of Captain Plum* and *The Wolf Hunters*, were published
in 1908. Thereafter, Curwood made his home in Owosso, traveled widely
in Canada, and devoted most of his time to fiction.[9]

In fiction and in life, he had always craved adventure. As a child in
Ohio, he once drifted into Lake Erie on a flimsy raft, only to be rescued
by a sailboat. Some years later, he was nearly arrested when he and a
friend embarked on a barnstorming tour selling "Infallible Blood Purifier,"

a homemade patent medicine. At one point, he lived for three weeks in a cabin in the swamps north of Owosso with an American Indian known as "Muskrat Joe."[10]

Even as a schoolboy, Curwood yearned to become an author. But his writerly education was tailored almost exclusively to the adventure genre. He pored over adventure stories in popular magazines and attempted to imitate their style. At first, he wrote in longhand on scraps of wrapping paper, which he bundled together as stories and novelettes. Later, his parents bought him a secondhand typewriter. The author's craft became an obsession. As a youth, he apparently wrote hundreds of thousands of words that went unpublished.

The young writer clearly viewed his own life as a nascent romantic fantasy. A life of letters would be a life of adventure and freedom: "While the minds of other school-children were learning... the hundred printed rules which a more intelligent system will some day send to the dump-heap of oblivion, I was building worlds, discovering continents, braving the hardships and perils of fanciful exploration."[11]

With the new century came a mania for "red-blooded" fiction, written by "men with the bark on" who had actually experienced the adventures they wrote about. Curwood needed a literary venue, a malleable landscape where he could meld real-life excursions with his romantic ideals.

He found it in Canada. In 1902, trapped in a newspaper job and an unhappy marriage, Curwood spent his free hours in the company of men who, like himself, craved the outdoor life. That year he met Malcolm V. MacInnes, Canada's immigration agent in Detroit. "Mac's" chief duty was to induce Americans to take up farming on the Canadian plains. Because the prairies lacked romance, the jovial MacInnes spent much of his time spinning stories of the Canadian North — the vast forests of the Canadian Shield, the glorious Rockies, and the tundra that stretched to the Arctic Circle and beyond. He introduced Curwood to a succession of picturesque and hardy Canadian visitors, from members of Parliament to rough-hewn trappers and lawmen of the Royal North West Mounted Police. Aware of a writer's promotional value, these Canadians enthralled Curwood with tales of the "almost unexplored wilderness of the north."[12]

With Mac's encouragement, Curwood began taking rail trips west from Thunder Bay to see the blossoming prairies for himself. But his real attraction was the northern forests, which he explored in 1904 during a trip to the Lake Nipigon region of Ontario. There he encountered, and shot, wild animals in abundance.[13] By 1908, when he was emerging as a novelist, Curwood was on the Canadian government's payroll, receiving $1,800 a year plus all expenses in exchange for unspecified boosterism on behalf of farming and tourism.[14] Soon he also was bartering his influence in exchange for land in several Canadian communities.[15] Though the land game enriched him, Curwood was never very interested in town building. His heart lay in the unpeopled regions of the North.

Soon Curwood was filling his tales with restless and brave characters who chafed at urban constraints. The hero of *Philip Steele* (1911) was a Chicago heir to millions, the toast of the country club set and its "hothouse varieties" of mindless amusement. But Steele had become a "hater of cities" and had renounced his fortune to ride with the Mounties in the wilderness. "Within himself he knew that he was unlike other men, that the blood in him was calling back to almost forgotten generations, when strong hearts and steady hands counted for manhood rather than stocks and bonds, and when romance and adventure were not quite dead."[16]

In *Philip Steele,* Curwood voiced considerable skepticism about the law, a theme he would emphasize for the rest of his career. Despite his admiration for some individual Mounties, he viewed their leaders as vaguely sinister, incapable of distinguishing true outlaws from those who had killed to defend their lives or honor. "Should I stand by and be shot like an animal just because it's the law that's doing it?" demands one character in *Steele.*[17] Justice and fair play, Curwood implied, were matters to be adjudicated by individual men, not by distant and unfeeling bureaucracies. The novel won him no friends in Ottawa. When a movie based on *Philip Steele* was released, it was banned in Canada.[18]

God's Country and the Bear

From his boyhood, Curwood possessed a streak of religiosity. In keeping with his temperament, the impulse was deeply personal and often floridly romantic. As a boy in Ohio he was taken to a Christian revival meeting. To the astonishment of his school chums, he fell into paroxysms of ecstasy and bounded onto the stage to proclaim his conversion. He recalled that a beautiful angel with long, flowing hair followed him home across the farm fields that night. His faith ebbed after several beatings from schoolyard bullies. "Slowly but surely I was licked back into mental health again."[19]

As a young man, Curwood formulated a new creed. True religion, he believed, was a matter of personal conviction based on the immanence of God. To believe in God, one merely had to look at nature. The glory of creation revealed the Creator. But civilization had corrupted religion. Human beings — including, apparently, the disciples of Christ — had obscured the truth with layers of interpretation, rules, and hierarchy. As with the law, Curwood was dubious of religious belief that was dictated by one's "betters." Just as true justice lay in the breast of the individual, so did the idea of God. Jesus had been a great "lover of nature" who had shed his "egoism" and recognized the dignity of all life. His followers had corrupted his teachings, twisting them into "a faith written in parables and riddles."[20] Curwood rejected the central tenet of Christianity: that every human being had fallen and needed redemption through Christ. Christianity, in this sense, ran counter to Curwood's romantic inclinations. It contradicted his stubborn belief that people (at least some of them) were capable of complete goodness.

Curwood's theology was compelling on paper but limited in its applicability. If God existed in nature, then God could be seen most clearly in places where nature was unspoiled. As expressed in the titles of several of Curwood's books, the wilderness was "God's Country." God dwelled where the vast majority of people did not. "We have no churches," one wilderness character declares in a 1911 Curwood story. "Our God is about us, in the forests, the swamps, the night play of the aurora."[21] Foreshadowing his experience with conservation politics, Curwood spurned organized religion as cumbersome and obfuscating. He failed to recognize that religion meant more than personal experience, that it also represented the idea of how to live in community. He was incapable of bringing his "pantheistic heart" to bear on human ethics.[22]

Curwood's ideas of God and nature were always intertwined, but they would be indelibly fused on a fateful day in 1914. That was the day when, alone and unarmed in the wilds of British Columbia, Curwood came face-to-face with "Thor," a 1,200-pound grizzly bear. In 1916 Curwood would immortalize the bruin in a novel, *The Grizzly King,* which he insisted was an accurate accounting of the real adventure.[23] Later he would describe the experience, in starkly biblical terms, as a journey to a mountaintop.

The hero of the novel is Jim Langdon, a writer of woodland adventure stories. His wilderness guide is Bruce Otto (named for Curwood's real-life guide, Jack Otto). The two venture into the mountains, where they track and shoot Thor. The giant bear is wounded. The adventurers give chase for several days.

At one point Langdon ventures from camp alone. He suddenly encounters the wounded Thor. The bear rears up in terrible fury — then decides he cannot attack this tiny and helpless being. "Thor was not, like man, a murderer." The grizzly ambles off, and Langdon is converted to bear worship. "You great big god of a bear!" he gasps. "You — you monster with a heart bigger than man!"[24]

One of Thor's keenest instincts is that human beings are invaders in the wilderness. The "man-smell" burns in his nostrils like a "plague."[25] Did this mean that God dwelled in a place where man did not belong? Did it mean, ultimately, that humans were defiling the Earth with their presence? Curwood never addressed these questions. At most, he recognized the spiritual and intellectual quagmire they entailed, and decided not to press them to their full conclusions.

Into the Fray

It is difficult to tell exactly what prompted Curwood to jump into Michigan conservation work. Perhaps it was the success of his novels, which boomed in popularity with *The River's End* in 1919. With fame and fortune assured, the author may have felt free to cast about in other fields. More likely, conservation was an escape from the workaday pressures of literary life and moviemaking.[26] Conservation appears to have given him an opportunity to bring his ripening personal creed to bear on public life. It is likely, too, that Curwood conceived of conservation work in unambiguous, individualistic, heroic terms — in sharp contrast to the maddening demands of literary celebrity.

Curwood the conservationist made a spectacular public debut. On a Friday evening two weeks before Christmas, 1921, he leaped onto the stage at an Elks Club in Flint, Michigan, before an eager audience of eight hundred sportsmen. Speaking fluidly and forcefully, he leveled a number of charges against Michigan's conservation apparatus. Incompetent men, he claimed, had been appointed as game and fire wardens in exchange for political favors. Because wardens were drawn from local communities, they often winked at fish and game violations by friends and neighbors. Concerned only with cronyism, state officials were standing idly as forest fires ravaged the cutover. The chief problem, as Curwood saw it, was the lax attitude of Michigan's new governor, Republican Alex J. Groesbeck, and his conservation director, John Baird. Curwood's audience sat in rapt attention through a speech exceeding two hours: "At times when he paused and announced he was about to spring some new sensation the big room was so still one could hear the proverbial pin drop."[27]

What was the impetus for Curwood's new crusade? He seems to have been scarcely aware of conservation's legal-economic and organizational revolution then in progress. Instead, he had gathered his evidence first-hand, by walking through the denuded forest lands of northern Michigan. By the middle of 1921 he had become "indignant and disgusted" at the situation.[28] He had seen the devastation wrought by fire; he had detected a general decline in the abundance of game animals and birds. He had heard tales of corruption, laziness, or incompetence among fire and game wardens. It was simple enough, from there, to fault "politics" for the sorry state of Michigan's natural resources.

To the extent that his ethos was worldly at all, it harked back to the Progressive Era: animals were disappearing and forests were burning because bad men were in power. The solution was to replace bad men with good ones, ridding the system of "politics" and emphasizing scientific expertise. The need for conservation was self-evident, and the people of Michigan — possessed of an instinctive virtue — would do the right thing if "politics" could be swept aside. Curwood outlined this faith to a member of the Conservation Commission: "I feel that to make Michigan the mighty state it should be, in beauty and health and natural resources, it is largely necessary to forget politics altogether. The average man is honest, and therefore, the average commonwealth is honest, and the commonwealth of Michigan will stand strongly behind honest endeavors."[29]

On the heels of the Flint meeting, Curwood issued a string of press releases and granted interviews decrying the destruction of Michigan's resources. Reporters enjoyed visiting his home in Owosso, where the floors were covered with bear rugs and the walls with original paintings of woodland scenes. Curwood was good company and "good copy"; the journalists scribbled dutifully as he fumed and mused. The corruption of the Groesbeck regime was a constant refrain. In his correspondence, Curwood sought signed affidavits — "red-hot sworn testimony," as he called it — attesting to the rottenness of John Baird's wardens.[30] Other than that, Curwood's specific enthusiasms appear to have been chosen almost at random, often at the prompting of others. Early in 1922, he latched on to a seasonal theme: "We are cutting at least a million Christmas Trees in Michigan each year.... [I]t is unwarranted destruction."[31]

His penchant for publicity did not go unnoticed. In Ann Arbor, P. S. Lovejoy watched the newspapers with a mixture of amusement and concern. Curwood, he felt, was a potential loose cannon. In a cautious and thoughtful letter sent just after the Flint meeting, Lovejoy sought a basis of understanding. Public sentiment, Lovejoy told Curwood, had to be cultivated in tandem with conservation methodology. Ousting bad politicians was not enough; a "new and more adequate machinery" of conservation had to be put in place.[32] Lovejoy's message had little impact. Without an entirely new political regime in Lansing, Curwood wrote in reply, "I feel that we ... will accomplish practically nothing." Asked to provide specifics of his program, Curwood provided none.[33] Lovejoy would try on several occasions to sell his land-use ideas to Curwood, to no avail. After one typically detailed Lovejoy missive, Curwood was unmoved and apparently uncomprehending: "I surely hope that a man who means as much as you do to conservation ... will not for a minute allow himself to be blinded or mislead [sic] by a lot of talk and theory while the thing we want is immediate action."[34]

Lovejoy's caution soon turned to disgust. On February 10, 1922, he took the morning train to Owosso and met Curwood for the first time. After a strategy session, the two proceeded to Flint, where they both were to speak to another sportsmen's gathering. With tensions between Curwood and John Baird at fever pitch, Lovejoy had cautioned Curwood not to drag him into any personal or political fracas. But he was nonetheless impressed by Curwood's publicity "machinery" and eager to use it.[35] By Lovejoy's account, the evening was an "abortion," with Curwood hogging the spotlight and the attention of reporters. "He seems to be sincere but if he really knows anything about 'conservation, propagation and reforestration' I didn't discover it," Lovejoy complained.[36] In

time, Lovejoy would acknowledge that Curwood had gotten the attention of Lansing politicians by "beating them over the head."[37] But he
would never take the author into his confidence.

In the meantime, Curwood had run squarely into one of the contradictions in his own logic: if he was calling for the ouster of Governor
Groesbeck, he had to advocate someone to take his place. As much as he
claimed to abhor "politics," such entanglement could hardly be avoided.
At the meeting with Lovejoy in Flint, Curwood had impulsively raised
his glass in a toast to industrialist Charles Stewart Mott, hailing him as
"Michigan's next governor." The toast made headlines and angered other
potential candidates, especially Fred W. Green of Ionia, a furniture man-
ufacturer and progressive Republican. Curwood, displaying more self-
awareness than usual, admitted that the gaffe had proved him "a rotten
politician."[38]

He was probably even worse at it than he knew. Mott decided not to
run. Curwood eventually endorsed the Democratic nominee, corporate
lawyer Alva M. Cummins, in the race against Groesbeck. In correspondence and press interviews, Curwood boosted Cummins as a conservation candidate. Only after goading by Lovejoy did Curwood contact
Cummins to seek his actual views on the subject. Cummins's answer
was a study in equivocation:

> I want to assure you that irrespective of all politics, it is my desire to see
> the wild life and forests of Michigan cared for as they should be. . . .
>   I must confess, however, that I have not given . . . such attention to the
> details of the proper conservation policy to justify me in committing
> myself definitely to particular proposals. . . . I am assuming, however,
> that policies favored by those who have given it careful consideration are
> apt to be right.[39]

Despite Cummins's bland reply, Curwood still insisted that the candidate was "100 percent" for conservation.

By this time Curwood clearly fancied himself a kingmaker. But the
illusion soon fell apart. Just before the 1922 election, he bragged to forester
Filibert Roth that "the time is coming when the forming of a conservation commission will be largely up to me."[40] Then Alva Cummins lost
badly; Groesbeck would be governor until the end of 1926.[41] With his
enemy firmly entrenched, Curwood wearied of the fight. For the next
few years, his conservation zeal would take other paths, including active involvement in the new Izaak Walton League. Here, as before, he

would be vocal, flamboyant, impulsive, often imprudent — and, in a way that is difficult to measure, highly influential.[42]

A World of His Own

In 1922–23, the citizens of Owosso witnessed the construction of a most unusual workplace. Overlooking the Shiawassee River near the city's downtown, it was a replica of a gatehouse to a Norman castle. Its walls were sixteen inches thick. The interior was dominated by an impressive "great room," with a huge fireplace and heavy draperies. Narrow steps led to the basement, which the proprietor liked to call "the dungeon-room." Anyone seeking entry to the building on weekday mornings would encounter a sign:

HALT!

Mr. Curwood
cannot be interrupted now.
He is doing original work.
Please Call Again.[43]

Architectural modernism clearly was not to Curwood's taste. "Curwood Castle," as he called his new studio, was a throwback to the days of medieval chivalry.

The new edifice said much about his life. Typing away at his desk in the castle tower, Curwood in the 1920s constructed more best-selling tales of the wilderness. As time wore on, the critics bemoaned the sameness of his novels. So he shifted his venue to historical romance. Instead of finding refuge in the woods, he found it in the past. Yet the song was the same. History, like the wilderness, was a landscape that he could populate with strong men, virtuous women, and unambiguous tales of individual heroism, unburdened by the constraints of modernity. His friend and editor Ray Long, a city man who viewed Curwood as his alter ego, put it aptly: "[T]here is a fine strain of sincerity in James Oliver Curwood that does not vary with varying conditions."[44]

Curwood's unease with modernity is seen most plainly in his novel *A Gentleman of Courage* (1924). Set on the north shore of Lake Superior, it tells the typical Curwood story of masculine bravery, feminine virtue, and the triumph of good over evil — "rather a standardized product," one critic yawned.[45] Yet it is unusual for its setting. Geographically, the story takes place closer to "civilization" than any other Curwood novel

Curwood Castle, the author's studio in Owosso, Michigan. Photograph courtesy of State Archives of Michigan.

since *The Courage of Captain Plum* sixteen years before.[46] In it, Curwood tells the story of a fixed human settlement. But in tying himself to one place, he betrays a severe anxiety over the workability of several human institutions: organized religion, the law, large-scale capitalism, and, by implication at least, even conservation. As opposed to his wanderers in the wilderness, who are faintly plausible, Curwood's village-dwellers have a sense of unreality about them. Though Curwood certainly did not intend it so, *A Gentleman of Courage* reads almost like a utopian fantasy.[47]

The action is set in the tiny community of Five Fingers, Ontario, named for the inlets of Lake Superior that reach toward the town. Five Fingers is a parklike little settlement, populated by a few dozen jolly people who are mostly French-Canadian and nominally Roman Catholic. Its economic sustenance comes from a communal sawmill. The town's only connection to the outside world is a black tugboat. Aside from the tug — an "ugly thing" that "did not belong at Five Fingers" — the community is entirely self-sufficient and harmonious.[48]

Into this setting comes Peter McRae, a boy who has been wandering in the forest with his father, Donald. Donald McRae is wanted by the provincial police for the slaying of a man in self-defense. Under hot pursuit, Donald sends Peter to Five Fingers, where he is taken in like a son.

Donald McRae, meanwhile, is chased through the wilderness by police for several years and driven to madness.

At Five Fingers, young Peter's sole enemy is Aleck Curry, the son of the tugboat captain. Aleck — the emblem of outside capital and corruption — is the only person at Five Fingers who is not physically attractive; he has a "large, coarse face" and close-set beady eyes.[49] Aleck's father has gotten rich from timber contracts granted by corrupt politicians. Aleck's uncle, who is politically connected, becomes commissioner of the provincial police. As Aleck grows into manhood, he becomes a bully on behalf of the government — by becoming a police officer. It is Aleck who hounds Donald McRae to the point of insanity; it is Aleck who, at the end of the novel, falls from a cliff at Five Fingers to a well-deserved death.

The lakeside paradise is saved, but what strikes the reader is its extreme fragility. Aleck Curry possesses the ability to "destroy everything" that makes Five Fingers "a fit place to live in."[50] The law is bluntly omnipotent, with little capability for distinguishing between criminals and honest men. A policeman's badge can be a refuge for "blood-sucking ferrets," one character declares.[51] At Five Fingers, order and justice spring from the hearts of the people, not from the dictates of some outside force.

Conservation does not fare much better. Pierre Gourdon, one of Five Fingers' hardy founders, regards every tree as a "word of God"; he mourns the sacrifice of timber in the community's "droning, merciless" sawmill but does not voice his lament aloud.[52] Apparently, the only thing saving the forest is the smallness of the community; Five Fingers can live in balance with nature only as long as it is tiny and isolated. Clearly, its worst possible fate would be to grow.

Laid bare in the novel, Curwood's theology looks vulnerable, as well. If Five Fingers is "God's Country," the rest of the world must be something less. People "buried in the darkness of cities" cannot hear the birds sing, Donald McRae tells Peter. Because the birds are animated by God, city-dwellers are cut off from the Creator's voice. "Always think kindly of people in the cities, Peter. They are unfortunate."[53] His words are echoed by Father Albanel, a wandering clergyman who has "no settled church" other than "ten thousand square miles of forest land."[54] Father Albanel preaches an evanescent creed of nature worship, without hierarchy, rules, traditions, or fixed beliefs, except a sort of hazy virtue to be found in the woods. His followers in Five Fingers are "free and easy and refreshingly happy," but there is no sense that his gospel would play in the city, where the handiwork of God is not self-evident.[55]

Egged on by editors and critics, Curwood in the 1920s sought a formula that reconciled modernity and romance. He was only partly successful. "Just as soon as I get out of the northland [as a setting for fiction] I lose my love for work," he confessed to Ray Long.[56] In *The Ancient Highway* (1925), Curwood went so far as to place much of the action in a city.[57] But the city in question is Quebec, whose crumbling fortresses offer the hero, Clifton Brant, an avenue of escape through history. Brant retraces the migrations of his heroic ancestors, who included an Indian princess. He meets up with Gaspard St. Ives, who might seem a quintessentially modern character: he is managing a huge tract of timber in the Quebec wilderness. But Gaspard, it turns out, is a man very much after Curwood's own heart. He lives in an ancient castle, because the stone walls invoke for him "something of the idealism and glory and thrill of a wonderful past." He "loves trees more than he does people."[58] Gaspard eschews the cool rationality of land economics; instead he is practicing scientific forestry because it recalls for him images of history and tradition.

In *The Ancient Highway,* Curwood revealed his limits as a chronicler of modernity. He could abide civilization only if it conjured half-accurate images of people long dead and events long past. As with Zane Grey, the cowboy writer who was his constant rival on the best-seller lists, Curwood's works constituted a pure mythic space, unbounded by limits of "historical or conceptual consistency."[59] In a Curwood novel, the setting was merely a backdrop for a formulaic plot that expressed the author's antimodernist longings.

If anything, those feelings only deepened as Curwood sought to retune his fiction for the times. He was deeply apprehensive about the workability of almost all organized solutions to problems of human existence. Truth in nature was self-evident, as were the means of nature's protection. Above all, there was good in the world and there was evil, and any reasonable person could see the difference between them. To fail triumphantly was better than to compromise. With such firm and dangerous beliefs, Curwood's existence entered its final chapter.

The Perils of Power

In November 1926, Fred Green won the Michigan governorship. At the end of December he named Leigh J. Young, a University of Michigan forestry professor, as head of the Conservation Department. And on January 4, 1927, he named James Oliver Curwood to the Conservation Commission, expressing "great happiness" over the author's potential as a policy maker.[60] Three of Curwood's fellow commissioners also were professional writers: novelist Harold Titus; Lee Smits, outdoor editor of the *Detroit Times;* and Norman Hill, editor of the Sault Sainte Marie *Evening News.* The other three were businessmen: Philip Schumacher of Ann Arbor, W. H. Loutit of Grand Haven, and Howard Bloomer of Detroit. Bloomer was the only holdover from the Groesbeck-Baird days.[61]

The Conservation Commission was a product of Governor Groesbeck's 1921 administrative overhaul, which had streamlined government by consolidating dozens of obscure boards and commissions in larger, thematic policy panels. The Conservation Commission shaped policy for the Conservation Department and made recommendations to the legislature. It had the power to set hunting and fishing seasons, but usually delegated these politically sensitive matters to lawmakers.[62]

Curwood assumed his post with his customary zeal. The new commission met on January 19 and appointed Curwood to head its committee on fish and game laws. Curwood was delighted with the appointment. Regrowing the forest would take decades, but the protection of wildlife was relatively simple: to save the remaining stocks of fish and game, Curwood believed, all that was necessary was to restrict their harvest. When he had scarcely warmed his chair at that first meeting, Curwood proposed that the killing of young buck deer (known as "spikehorns") be banned. The panel took the proposal under advisement.[63]

Back in Owosso, Curwood decided that the conservation emergency could not wait. By late January he had drafted a bold plan calling for slashing the deer season from fifteen days to seven, doubling deer license fees, creating "sanctuaries" on trout streams where no fishing would be allowed, and shortening seasons on rabbits and other small game. The most inflammatory part of the plan was a three-year moratorium on the hunting of ruffed grouse, a staple for northern Michigan sportsmen. He sent each commissioner a copy. Already he had identified himself, Titus, and Lee Smits as the true "outdoor men" on the commission.[64] And he feared — somewhat ironically, given his supposed support for scientific expertise — that Conservation Director Young might be too deliberative to meet the emergency. Young, Curwood confided to

Governor Green, "has a long established background of acedemic [sic] training and opinion which tends to scientific and investigatory instead of heroic action."[65]

Curwood's fears soon materialized. At the February meeting, his fellow commissioners sent several of his proposals — including the grouse-hunting ban — to the legislature with a favorable recommendation. But dissension was rising. Harold Titus and Norman Hill favored "a more liberal policy towards the sportsmen," a reporter noted.[66] And legislators did not take the "heroic" course that Curwood had recommended. They passed a watered-down fish and game package containing none of the shorter seasons the commission had called for.[67]

By April, Leigh Young had learned of Curwood's disapproval, and the two men had a falling-out. Rumors swirled that either of them might quit at any moment.[68] William B. Mershon, a Saginaw lumber merchant who had become one of Michigan's leading conservationists, urged Young to hold fast under Curwood's pressure. "Don't you let him drive you off the roost," Mershon wrote. "He is simply an impractical trouble breeder."[69] Threats of resignation were all the rage that spring. In May, Curwood and Smits both talked of quitting after they quarreled over the commission's firings of high-ranking game and forest-fire officials. But the rift apparently was patched.[70]

The personnel battles had deepened Curwood's conviction that Young was too much the deliberative academic and too little the man of action. "The cry on Mr. Young's part to 'Go slow, go slow,' is all wrong," he complained to Harold Titus. Curwood's impatience was compounding itself. He hinted to Titus that the commission should restrict hunting seasons on its own.[71] At the June meeting he won a partial victory: the commissioners voted to ban grouse hunting for a year.[72]

In July, Curwood lost his patience. He again placed his program before the commissioners and asked them to act where legislators had not. But commissioners balked at the demand. The Conservation Department had distributed twenty thousand questionnaires asking sportsmen's opinions on the Curwood plan. Only about two thousand had been returned by July, but they did show a majority of respondents in favor of Curwood's ideas. Curwood called for a quick vote. But Commissioner Loutit advised that, because only 10 percent of the questionnaires had been returned, action should be delayed.

At that point, Curwood blew up. In a tirade that a reporter likened to "the storm scene by the Moscow Opera company," he blasted commissioners for their "fear to act."[73] He pointed out that most respondents agreed with his plans. Curiously, he then argued that the responses didn't matter, because the questionnaires had been distributed haphazardly and were subject to fraud. Curwood claimed, apparently without proof, that a notorious poacher had filled out three of the forms, each time opposing Curwood's plan to protect "spikehorn" bucks. "If you want to take the word of that bootlegger against mine, it is time I got off this commission," Curwood fumed.[74]

The "storm scene" succeeded, at least in the short run. Commissioners refused to enact the controversial "spikehorn" plan, but they agreed to reduce the taking of deer, rabbits, muskrats, and squirrels.[75] Weary of the battle, Curwood retreated to his rustic lodge along the Au Sable River in northern lower Michigan.

His agitated mind found no peace in the forest. He wrote conservation polemics for the newspapers and brooded over his own conservation ethos, which was rife with contradictions. Trying to mend fences with fellow commissioners, he declared that the commission was not "a political organization." But commissioners needed the support of the public if they were to gather their courage and take action. Sportsmen had only themselves to blame if the commission misjudged public sentiment, Curwood wrote.[76] At the same time, he believed that the commission's mandate was far wider than satisfying the blood lust of hunters and anglers. Curwood was not entirely opposed to hunting; he always left room for the shooting of game to meet human needs such as food. And he remained an occasional fisherman himself. But he never fully articulated the difference between sportsmanship and butchery. From this generalized moral revulsion against blood sports, Curwood leapt to the conclusion that sportsmen's wishes sprang from bad motives. When Lee Smits tried to cheer him up after the July meeting, Curwood was unmoved. Every gain won for wildlife, Curwood wrote, "has been achieved over the dead bodies of Mr. Titus and Mr. Loutit. In my opinion these two men follow the destructive tendencies of the last commission and eternally have their ears to the ground to see what the hunters and fishermen are going to say. . . . In every instance they listen for the voice of the killer."[77]

Returning to Lansing for the last time in August, Curwood mustered support for the "spikehorn" measure. With the four votes for passage apparently secure, the Conservation Commission laid the matter over for a month.[78] Curwood went home to Owosso feeling gloomy and feverish.

For years his health had been precarious. In his writing he had bragged that he would live to be one hundred, thanks to exercise and the avoidance of red meat. But like many of his boasts, this was an elaborate fantasy that belied the facts. For a decade now, he had overextended himself. He had written a novel a year, sometimes two; he had produced films and articles and stories; he had poured his limited energies into crusades that brought much frustration and little joy. The functioning of his body mirrored all the excesses and vulnerabilities of his restless mind.

In August 1927, the pace caught up with him. A few months earlier he had suffered a leg wound — perhaps a sting or snake bite — while wading in a swamp in Florida. The pain had lingered, despite treatment. Now it flared into a dangerous infection. Within a few days, he was delirious. His daughter supplied a blood transfusion; two doctors from Detroit arrived by airplane. But the effort was to no avail. Curwood died shortly before midnight on August 13.[79]

Meeting on September 7, the Conservation Commission wrote an ironic remembrance for its late member. The "spikehorn" measure was set aside. In addition, the commissioners rescinded earlier orders that had restricted hunting of rabbits and squirrels. They also lifted a two-year-old restriction on fish spearing in southern Michigan. In the words of one reporter, "The whole foundation of Curwood's conservation program . . . was torn to pieces with almost vicious deliberateness."[80] Afterward, the commission unanimously approved a resolution mourning Curwood's passing and expressing its "keenest appreciation" for his efforts.[81]

Epitaph: *Green Timber*

In 1930, a Curwood novel was published posthumously. Titled *Green Timber*, it had lain uncompleted at the time of his death. Curwood had finished six chapters and had outlined the remainder. A Curwood novel — even one half finished — was too valuable to discard. So the manuscript and Curwood's notes had been farmed out to Dorothea A. Bryant, who also was updating and expanding Curwood's autobiography.[82]

Green Timber is not set in the wilderness. It is set in northern Michigan's cutover country. For Curwood, this was an extraordinary concession. The cutover held only faint appeal to Curwood; he had always invested his dreams in the "unspoiled" reaches of Canada and Alaska instead. The cutover smacked of politics and corruption and waste, of depleted forests and faded glories. But with the real-world problems of conservation pressing in on him, perhaps Curwood had felt the need to look at Michigan's situation head-on.

The novel tells the story of a petty criminal, Allan Campbell, who is being hunted by gangsters in Detroit. Campbell escapes to the cutover. There he undergoes a sort of moral conversion, becoming a "sturdy frontiersman" and a crusading conservationist.[83] In the end, he defeats his gangland enemies in a gunfight along the Au Sable River.

In Curwood's view, the cutover is a "second-growth wilderness," a land that is thriving because it has been abandoned. Nature—not law or science or publicity or politics—is the sole agent of Curwood's cutover revival. Natural forces are turning the cutover into "one of the garden spots of the earth."[84] Because human beings have taken their destructiveness elsewhere, the cutover has become God's Country, a "land that men—but not God" have forgotten.[85] The absence of law allows Curwood, and his redeemed gangster-hero, to reclaim the cutover as a romantic landscape. At the end of the novel, with bullets flying all around him, Allan Campbell finds time for a speech about the region's heroic possibilities. With the law far away, he proclaims, "wilderness" folks must take matters into their own hands: "In the cutover lands we are masters, for the law has abandoned us."[86]

Curwood's conception of the cutover was entirely wrong. In real life, the cutover bloomed because law was *applied* to it, not because it was withdrawn. More than a work of nature, the remaking of the forest involved the conscious projection of human values onto the landscape through the machinery of the state. Much of this machinery—including land-economic surveys, timber-taxation laws, and organized fire protection—had come into being during the half dozen years when Curwood was active in conservation. Under the tutelage of persons wiser than himself, such as Filibert Roth and P. S. Lovejoy, he had paid occasional tribute to these mechanisms. But as *Green Timber* showed, he had never abandoned his faith in the romantic hero or his skepticism about the workability of the law. He could learn to love the cutover only by viewing it as a landscape of pure individualism—something it absolutely was not.

By any assessment, Curwood was a one-dimensional thinker. But his single dimension was spectacular, even admirable, for all its blind zeal. As an antimodernist policy maker—a glaring oxymoron, though Curwood didn't realize it—he rejected the bureaucratic rationalization of his cherished woods. He identified the human heart as the locus of all human progress. He never varied his message with the audience and never distinguished fiction from fact, myth from reality, or individualism from the imperatives of modernity. In Curwood's eyes, there was only one story, and he probably conceived of himself as its hero.

Yet even people who condemned Curwood as a fool (and there were many of them) were fascinated by his hold on the public imagination. Curwood's presence was a reminder that not everything could be quantified and that the creation of a rationalized natural environment would depend on sentiments that were nostalgic, mythical, even irrational. In years to come, his individualistic creed would flower through organized means, most notably in the wilderness movement.[87] Like his valiant men of the North Country, the keeper of Curwood Castle had not labored in vain.

CHAPTER SIX

The Production of Leisure

Forest Recreation and the Boundaries of Nature

> We have great machinery to produce joy. . . . I do, however, insist that
> no other organized joy has values comparable to the joys of the out-
> of-doors.
>
> — HERBERT HOOVER, 1927

On January 14, 1922, fifty-four professional men met in Chicago and
formed what would become the largest and most influential conservation group of the decade. The Izaak Walton League of America, named
for the seventeenth-century author of *The Compleat Angler*, was a national club for fishermen and hunters. Though sportsmen's clubs were
nothing new, the "Ike Waltons" promoted a fresh agenda: jumping into
the realm of politics, they aggressively championed the artificial propagation of fish and game, so that sporting men might have something to
catch or to kill.[1]

Beneath this methodological prescription lay a novel, even radical,
social rationale: If Americans were to preserve any vestige of the sporting experience that had defined the frontier, they would have to do it
through organization. In an impassioned editorial in the league's new
magazine, cowboy novelist Zane Grey endorsed collective action as the
only means of preserving "the love of outdoors that made our pioneers
such great men." It was a stunning irony. The author whose works celebrated the lone gunman now was implying that individualism, at least
in the arena of blood sports and their role in forging the American character, was obsolete and even dangerous.[2]

The Walton League's work was an outgrowth of a larger trend toward intensive resource management for the purpose of public leisure. Henry S. Graves, the nation's chief forester, had set the tone in 1920 by demanding quick action toward a national policy on outdoor recreation.[3] In so doing, he had gone far beyond Gifford Pinchot's belief that the government's forest existed to provide three essential goods: timber, flood control, and a brake against monopoly power and ruinous competition. Graves now asserted that collective action, through planning and scientific management, was the only way to satisfy the public's yearning for wild places. Having conquered the frontier a generation before, Americans now wanted to get back to it — "to seek the wholesome environment of the hills and forests and to refresh mind and body through the vigors of mountain and camp life."[4]

Through the actions of government and private citizens, reinforced by the mass media, the 1920s marked a watershed in the history of American outdoor life. For the first time, the encounter with nature was identified as a commodity that could be defined and administered — not just in response to public demand, but proactively, as a way of perpetuating values deemed important to American democracy. Forest recreation thus was added to the list of desirable outputs of a "continuously productive nature" to be had through centralized planning. Citizen groups such as the Walton League, recognizing the individual's interest in the maintenance of forest play-places, would become forceful apologists for government action in what formerly had been conceived of as an "untouched" landscape. The synergies would be commercial as well, with media and advertisers realizing that outdoor life could be linked with a vast array of consumer products and services. The rural playgrounds of America in the 1920s would become outposts of law, science, administration — and business.

In the process, the social argument for the forest's existence would change as well, going beyond questions of efficient production to those of consumption and quality of life. A decade that had started with talk of a "timber famine" would end quite differently. "People are no longer much concerned as to sources of building materials or of food," a forester noted in 1930. "They are becoming increasingly concerned as to sources of beauty, inspiration and recreation . . . under which the traditions and characteristics of American life can be perpetuated."[5]

The science of recreation, born of Progressive anxieties over the idle hours of immigrants, would be expanded in the 1920s to encompass the leisure activities of the growing middle class. Widening their focus from the city playground to the forest, recreation planners would evince two of the same tenets that had marked their earlier crusade: a faith that the "quality" of each person's recreational experience could be consciously shaped through management and a belief that constructive play was the best means of fitting people for productive work and citizenship.[6] Forest recreation planners believed that the healthful life out-of-doors could help steer Americans away from enjoyments that were essentially vicarious and artificial — such as motion pictures — and from others that were "vulgar in expression."[7] In this sense, the recreation movement of the 1920s retained elements of the Progressive impulse to guide and control individual behavior. Herbert Hoover, who seemed capable of applying the engineer's dispassionate method to any social problem, spoke of "organizing the production of leisure." If fishing, for example, was good for the soul — as Hoover emphatically believed it was — then government had an interest in perpetuating it, by controlling pollution and stocking hatchery-raised fish in lakes and streams.[8]

As Hoover and others recognized, the surging popularity of outdoor recreation in the 1920s stretched the natural environment to a breaking point. The symbol of the age was the automobile, which suddenly made even distant woods and waters accessible to the middle class. As the decade wore on, a new generation of professional managers realized the necessity of channeling the tourist hordes to protect the countryside from overuse and to safeguard private property rights. By 1930 they also had built the rudiments of a fish and game bureaucracy, dedicated to perpetuating wildlife for the benefit of the angler and shooter. Even with the development of these protective and propagative measures, some planners feared that footloose Americans might overrun their environment entirely. As late as 1933, for example, one observer in the Hoover administration predicted that "hunting and fishing in the near future may die out as popular sports . . . and be available only to those able to hold memberships in exclusive clubs" that would grow their own game and fish. That undemocratic fate could be averted only through "statesmanlike planning," the writer stated.[9]

Four case studies — automobile camping, game management, the growth of the Izaak Walton League, and the genesis of the wilderness movement — are offered here in support of the idea that forest recreation increasingly was restricted by certain boundaries during the 1920s and early 1930s. These boundaries were conceptual as well as physical.

They involved numerous limits on the range of human action in a formerly unregulated environment. At the same time, state and private
management often were intended to preserve some vestige of frontier
experience by protecting resources from the ruinous consequences of
unregulated use.

Auto Camping and the Challenge of Containment

In 1921, the U.S. Forest Service issued a small pamphlet inviting the public to camp in the national forests — "a place where you may feel at home
and enjoy yourself in your own way." The publication made clear that
forest camping was a casual and unregulated affair: "Camping is free and
generally requires no permit. You may choose your own camp ground
and help yourself to dead wood for fuel." Anyone with an automobile
was welcome to drive into the forest as far as he dared and pitch a tent.
The only admonishment was to be careful with fire.[10]

Automobile camping — known also as "autobumming," "flivverist-
ing," and "motor caravanning" — boomed in the early 1920s. The nation boasted 2.5 million miles of roadway in 1920, though nine-tenths
of that was characterized as "unimproved."[11] For some, that daunting
figure seems to have constituted a romantic challenge. In narrative accounts of camping trips, flat tires and rutted roads often were likened
to the hazards faced by the pioneers. A magazine piece titled "Living in
the Car" spoke not of deprivation, but of the grand adventure had by a
Buffalo man who spent four months on the road. By outfitting his 1920
Franklin touring car with detachable bunks, an outrigger tent and kitchen
equipment, he had transformed the auto into a "Locomotel."[12] An array of commercial products promised to turn every automobile into a
den of luxury. The "Burch Auto Bed and Tent," which attached to the
side of a car, was advertised as "lots better than a hotel. . . . [Y]ou get
up chuck full of vim and snap, ready for keener enjoyment of the plea-
sures of the great outdoors."[13]

Campers needed a place to stay, of course. Some simply pitched their
tents on private land, but the practice annoyed landowners as camping
became more popular. In the early 1920s many municipalities set up free
tourist camps, often including fresh water, picnic tables, communal fireplaces, even hot showers. The aim was to keep campers off private land,
and to cluster them near grocers, tourist attractions, and other sites where
they might spend money. When a Minneapolis man and his brother jour-

Automobile campers at Wabikon Lake near Laona, Wisconsin. Photograph courtesy of State Historical Society of Wisconsin, WHi (W6) 31528.

neyed west in 1921, they stayed at free camps almost every night during the six-week trip, incurring total expenses of just $242.50. The "gypsy" life of the tourist camps was casual and congenial, replicating the days of frontier sociability: "We had lots of company. Tourists everywhere; men, women, and children wore khaki, looked like hoboes; were dirty, dusty, and the men were often unshaved, but everywhere everybody was happy, cheerful, healthy, and all declared themselves to be having the time of their lives."[14]

Like most new social phenomena, the automobile boom gave rise to all sorts of florid predictions. Frank A. Waugh, who wrote frequently about landscape architecture and community development, celebrated the open road as a "national park" in itself. In 1920, Waugh estimated, there were five million private automobiles in America, being driven an average of seven thousand miles a year, with half of that devoted to pleasure trips. With three passengers per trip, the statistics on motor touring could make one positively giddy: "[F]ifteen million Americans use, but do not consume, fifty-two and a half billion miles of scenery every year."[15] Another writer predicted that auto camping could hold myriad social benefits: freed from the shackles of small-town isolation, the villagers of America could motor forth to mingle with city folks in tourist camps, with enlightenment accruing to both sides.[16]

A "Ford camping outfit" of the type popular in the 1920s. The cover of the trailer doubled as a small boat. Photograph courtesy of State Historical Society of Wisconsin, WHi (D483) 1211.

The euphoria would not last. Within a few years, magazine articles increasingly were referring to auto camping as a "problem" in need of a solution. The reason was the sheer number of auto tourists, which put unbearable pressure on both private and public facilities. From five million in 1920, the number of private autos soared to twelve million just four years later.[17] The swarm of campers raised problems of sanitation, trespass, traffic control, and forest fires. Eight million people visited the national forests in private cars in 1923, jeopardizing life and timber with their cigarette butts and campfires. In anticipation of a bigger onslaught in 1924, forest rangers were ordered to patrol the roadsides and sniff out firebugs.[18] The promoters of municipal camps, meanwhile, found their facilities being used not just by free-spending middle-class tourists, but by vagabonds who were making a permanent home on the road. One critic complained of "the automobile hobo," who invariably drove "a broken-down flivver" and lived off "the hospitality of others."[19] Free municipal camps were in decline by 1925, replaced by private or public camps that charged per-night fees as high as a dollar.[20]

The "problem" of camping was best dealt with through physical containment. By the mid-1920s, landscape architects and recreational planners had refined a design for the modern yet rustic auto campground.

Accessible by a loop road, it offered grassy meadows or forest clearings where motorists could drive in and pitch their tents. Dirt or stone fire pits minimized the forest fire hazard. Fresh water, garbage cans, and toilet facilities were centrally located. Clumps of trees, strategically placed, gave campers at least an illusion of isolation. Yet a salient feature of the new campgrounds was the possibility of social interaction. Some featured lodges where campers could warm themselves around a fireplace, tell stories (true or not) of outdoor life, or conduct religious services.

Both the landscape and the buildings reflected a naturalistic architectural style, with materials that replicated and harmonized with the natural features of the site. Campers wanted to escape the "artificial" and "cultivated effects" that spoke of "the routine of life," a Michigan landscape architect noted. But the naturalistic mode was hardly "natural" itself. It was, rather, a meticulously planned style that sought to "enhance and perfect the dominant type of beauty" of the existing landscape. The naturalistic campground featured eye-pleasing arrangements of native trees and plant species, and employed materials such as rough lumber and stone for construction of picnic tables, fireplaces, and other amenities. By coordinating each detail, the builder of a campground could achieve "a perfect interpretation of nature's character" — something better than nature itself.[21]

Equally important from the planner's perspective, the new campgrounds offered optimum control over camper behavior, while maintaining a veneer of rugged individualism. Clustered together, the camping sites could be patrolled by rangers, police, or private personnel to enforce standards of cleanliness, decorum, and especially fire prevention. Access could be controlled by user fees or permit systems, and "hobos" could be discouraged by rules limiting length of stays. Popular camping places, one forester wrote in 1924, actually had become seasonal cities, with all the problems of sanitation and law enforcement that city life entailed.[22] By 1927, another observer noted sourly, the average camper had to satisfy his yearnings in "well-patrolled and so-called fireproof camping districts" under the watchful eyes of "speed cops, forest rangers, park police and village constables."[23]

The net effect of this trend was to limit the average camper's options on where to pitch his tent, while increasing the amenities available to him at controlled camping sites. As late as 1920, a good deal of motor camping apparently had been done randomly on private land. "A few years ago," one man observed in 1924, "a motor camper could practically pitch a tent anywhere in the country districts, provided he looked and acted decently and asked permission." But because some motorists had treated the countryside as a "waste barrel," all that had changed. Now more lands were posted against trespass, and the tradition of country hospitality (never much of a burden in horse-and-buggy days) was giving way to chain-link fences.[24] Because of fire hazards, the Forest Service discouraged random camping in the woods and instead steered recreationists to the 1,500 public campgrounds it had laid out by 1925. Many of these were managed cooperatively with local civic clubs or chambers of commerce. A "certain amount of policing" was required at these sites because many of the "new class of visitor" were "city bred" and knew little of the outdoors, a report stated.[25] Private campgrounds increasingly found it profitable to be linked by membership organizations that published guidebooks listing amenities and directions for motorists. Motor-camping clubs held the additional, unspoken appeal of segregating campers by social class. The National Recreation Club of Boston, for example, catered to the refined camper who favored quiet enjoyment of sites off the beaten path. For five dollars a year, plus nightly fees, members got access to pastoral farm sites whose owners otherwise might have been hostile to motorists: "The man coming to him with a membership in a club... is a voucher to the farmer that his property will not be harmed."[26]

Campground design would be further solidified in 1932, with the issuance of a standard design policy by the Forest Service. The document was the brainchild of E. P. Meinecke, a noted plant pathologist, who had consulted with park planners and foresters since the mid-1920s. Meinecke had warned that cars and foot traffic were destroying vegetation in high-use areas, such as Sequoia National Park in California. If not strictly contained, the automobile was a veritable killer in the forest: it crushed young roots and dripped oil, which was deadly to plants. Meinecke's solution was to divide the campground into discrete and uniform "campsites," each with a cleared "garage spur" where an auto could be parked. Each site had its own fire pit, picnic table, and level spot where a tent could be pitched. Stone and log barriers — natural in appearance but very deliberately placed — would prevent motorists from wandering. Gone were the days when a motor-camper could choose his site by driving across a meadow or even into the forest itself. "Instead of permitting campers to do their own haphazard planning," Meinecke wrote, the experts would now do it for them. So far-reaching was Meinecke's vision that, for two decades afterward, designers would refer to such campground renovation as "Meineckizing."[27]

By placing great strain on natural resources, the auto camping boom had revealed the hazards of rampant individualism in a finite environment. Landscape architects, foresters, and planners tackled this problem by bringing law, science, and administration to bear on what formerly had been a matter of individual preference. By the late 1920s, camping had become a collective and somewhat regimented exercise, albeit one in which the individual could still find significant enjoyment. The camping experience had been tamed.

"Administering" Animals: The Rise of Game Management

In the 1920s, having grown rich as a Michigan lumber merchant, William B. Mershon began a second career as a conservationist and outdoor writer. Many of his tales involved reminiscing on the old days of hunting and fishing. Mershon was old enough to remember when (in the 1870s) it was common to see bears wandering into Saginaw, especially when the nearby forest was burning. Now the bears were mostly gone, as was the forest. Gone, too, were the spectacular wild creatures that bespoke Michigan's frontier heritage: the passenger pigeon, the arctic grayling, and the wild turkey. In 1922, Mershon wrote a wistful remembrance of turkey hunting in Michigan. He believed he had killed the last wild turkey in the state, a twenty-two-pounder that fell victim to his gun in 1895. "Radiant as an oriental jewel box," the bird had been stuffed for display in Mershon's home, a glass-eyed witness to a vanished era. As Mershon saw it, the noble old gobbler was "The Last of His Race."[28]

Mershon typified a lingering attitude circa 1920: that wild animals constituted a finite stock whose disappearance probably was inevitable. Because animals came from nature, it was thought that the tide of human civilization would overwhelm and destroy much of what nature could produce. As for Michigan's glorious turkeys, "it is too bad that they are gone forever," Mershon wrote, but the tone of his article was more nostalgic than regretful.[29]

If the defenders of wildlife were uncertain that they could halt its destruction, at least they were confident that they could slow it down.

Law, augmented by public opinion, was the chief means of protecting animals before 1920. Under American law, wild animals were the property of the public and had been accorded some state protection even in colonial days.[30] The nineteenth century saw public crusades against "market hunters," who shot wild game for commercial sale, and against recreational "game hogs" whose wanton killing was condemned as unsportsmanlike. Closed seasons and daily "bag limits" restricted the sport hunter's take, though enforcement was sometimes spotty. In the early twentieth century, the federal government had banned the sale of wild game and claimed jurisdiction for protecting migratory birds. From the Civil War onward, restrictions on harvest had been championed by sport hunters, at least by those who embraced the English model of gentlemanly conduct afield. By 1900, the value calculus for wild animals had begun to change, with the creatures being prized not so much for their meat as for their role in facilitating the sporting life out-of-doors.[31] In the eyes of Wisconsin's Aldo Leopold, however, the wildlife crusade had been essentially reactive: "In America the dominant idea, until about 1905, was to perpetuate, rather than to improve or create, hunting. It sought, by restrictive measures, to 'string out' the remnants of the virgin supply, and make it last as long as possible. Hunting was thought of as something which must eventually disappear."[32]

In the 1920s and '30s, Leopold and a handful of others would establish the intellectual, administrative, and scientific framework for a new approach to American game. They would augment "game protection" with "game management," taking a proactive role to increase populations of birds and animals. Instead of merely restricting hunters, they would focus on the crucial role of game habitat, cover, and food, all of which could be provided by human beings. They would begin to study the complex web of animal ecology, including natural cycles of plenty and scarcity and interdependence among species. By 1931, Leopold could write: "Both scientists and sportsmen now see that effective conservation requires, in addition to public sentiment and laws, a deliberate and purposeful manipulation of the factors determining productivity—the same kind of manipulation as is employed in forestry."[33] Game management thus fell under the rubric of continuous production, on two levels: managers would deliver a constant stream of birds and animals, which would facilitate a succession of enjoyable days afield for American sportsmen.

Leopold and colleagues also would grapple with the toughest social and political problem in game management: although wild animals were the collective property of the people, hunters could not give chase across private lands that were posted against trespass. Leopold, acutely conscious of hunting's place in frontier lore, wanted to preserve some semblance of frontier experience before it was too late. Like many men of his generation (he was born in Iowa in 1886), Leopold fondly remembered the days of free access to hunting, and he feared that the United States would become a vast "private game preserve" by 1940. Even if sportsmen could afford the fees charged to shoot on fenced acreage, they had an "ingrained repugnance" against "having their sport served to them in a spoon," Leopold wrote in 1919.[34] The solution was to cultivate game stocks on public lands, while negotiating agreements that might keep private acres open to free public hunting.

Aldo Leopold's peculiar genius lay in his ability to move fluidly between the details of daily conservation work and the esoteric task of defining a new environmental ethic, always using the fruits of one pursuit to inform the other. A prolific and gifted writer, he constantly sought to explain his work to a wide array of publics, from citizen conservationists to specialized professionals. As a federal forester, independent game consultant, and University of Wisconsin professor, he not only advanced a new system of game management, but explored the system's ironies and limitations as well.[35]

The chief tension in game management was this: because game stocks and accessible land both were declining in the 1920s, game managers had to take affirmative measures to replicate the hunting experience of pioneer days. By manipulating the environment—say, by establishing public hunting grounds or planting food that wild animals liked to eat— they might increase the number of animals available for hunters to kill. But "excessive manipulation of environment tends to artificialize sport, and thus destroy the very recreational values which the conservation movement seeks to retain," Leopold admitted in 1930.[36] "Artificialization" might be distasteful, but it was utterly necessary in a pressured environment that could not meet hunters' demands without human help. "Every head of wild life still alive in this country is already artificialized, in that its existence is conditioned by economic forces," Leopold wrote, with a certain sense of resignation, in 1933.[37] Although game man-

Aldo Leopold, about 1929. Leopold was among the most prominent apologists for the "artificialization" of natural resources. Photograph courtesy of University of Wisconsin–Madison Archives.

agers had an affirmative duty to control nature, they could do it with a certain finesse that would camouflage the human element, and thus preserve at least the appearance of rugged sport. The ducks or deer propagated for hunters' pleasure might be "artificialized," but the experience of pursuing them needn't be entirely sterile. Game management, in this conception, would be a naturalistic practice not unlike landscape architecture, using human ingenuity to replicate and amplify the workings of nature — in the process, creating an "imitation" of nature that would serve human needs better than nature itself.

Under this framework, the natural phenomena of animal life gradually would fall under the influence of human law, science, and administration. Leopold knew, for example, that every healthy wildlife population produced an annual "increment," or increase in population, that could be harvested without decimating the animals' breeding stock. Through active management, that annual mortality could be achieved through hunting, rather than through starvation or killing by animal predators. In other words, even the sobering reality of death in the wild could be channeled to serve distinctly human ends. A forester lacking Leopold's eloquence put it bluntly in 1928: in an "enlightened age," it was simply "absurd" to allow animals to starve to death, instead of seeing to it that they were shot by hunters.[38] In Michigan, Leopold's friend P. S. Lovejoy was establishing a network of state game refuges. Closed to hunters, these state-owned parcels would be intensively managed to produce game that would migrate to nearby public hunting grounds, where it could be slain by sportsmen. If strategically located, the refuges might pour forth "a ready and steady spread of game in all directions."[39] Management, it seemed, could make the animal kingdom continuously productive.[40]

The evolving practice of game management got a dramatic demonstration in Michigan, with a joint experiment begun by the Izaak Walton League and state conservation authorities in 1930. The Williamston National Project undertook the management of an entire rural township, about nineteen thousand acres, not far from Lansing. Farmers in the township were upset over hunter trespass and bad behavior. Such complaints had prompted the legislature, in 1927, to approve a strict trespass law targeted specifically at sportsmen.[41] At the same time, the Williamston farmers — many of whom were hunters themselves — were open to suggestions as to maintaining the tradition of free access, while at the same time getting state help with rearing small game on their land. Leopold had an advisory role and an active interest in the project, believing that it could point to ways of managing animals and the hunters who pursued them.

The most striking aspect of the Williamston project was the intensity of the management effort, which sought to bring every acre, every animal, and every sportsman within the grasp of a centralized oversight team. Access to hunting was controlled through a permit system. Each farmer got a prescribed number of yellow entry tags, which he could hand out to sportsmen as he saw fit each day. No fees were charged, but the system kept hunters on their "best behavior," one landowner said. Just in case, a "central committee" stood by to prosecute offenders, so farmers would not have to press charges on their own.[42] The management of the animals was no less intense. After the township was mapped for soil characteristics and available game cover, plans were drawn up for "improving cover, feeding and nesting conditions." By January of 1931, sixty-seven artificial feeding stations had been set up with help from Michigan State College. Pheasants, rabbits, and squirrels were helping themselves to two tons of feed, with more on the way. "The wild creatures of Williamston township must think that times are pretty good, and that this part of Michigan is a good place to spend the winter," wrote a project organizer, Earl C. Doyle of the Izaak Walton League. The aim was to encourage natural reproduction and "to let Nature do her work better" through human aid.[43] The upshot of the project, Doyle wrote, was to achieve a far greater population of shootable game than could ever be accomplished by "merely turning birds loose to shift for themselves."[44] Whether the pheasants knew it or not, theirs was no longer a laissez-faire landscape.

In the midst of this unprecedented management effort, Doyle also could write — apparently with no sense of irony — that the goal of the Williamston project was identical to that of the Izaak Walton League: "To Restore the America of Our Ancestors."[45]

To be fair, Earl Doyle was a deeply thoughtful man who, if prodded, certainly would have admitted to the twists of logic inherent in modern natural-resource management. In a memo written in 1931, he acknowledged these tensions at least obliquely. Izaak Waltonism, he asserted, was a "cultural mission" whose aim was to promote individual fulfillment through organized means. "To fish, to hunt, to build campfires and enjoy the great outdoors will not be mere sport. It will be a form of self-expression, a chance for unfolding of personality." The same "machine civilization" that had created so much free time for Americans now threatened to overrun the countryside. The imperative for managers was to build an administrative bulwark that would "serve the individual."[46] In game management, the immediate choice was clear: either a drift toward depletion of game stocks and sterile privatization of the sporting experience, or the mastery of propagation through "artificialization," which would provide at least a touchstone for the hunting legacy of the vanished past.

The "Ike Waltons" and the Commodification of Outdoor Adventure

Advocates of wildlife management had no better friend than the Izaak Walton League. Its fundamental mission, to use Aldo Leopold's phrase, was to promote the "purposeful manipulation" of the conditions under which fish and game lived, and thus to increase the sporting opportunities for anglers and hunters. This meant controlling water pollution, setting aside wildlife refuges and public hunting grounds, and campaigning against drainage of wetlands. It also meant championing the artificial propagation of wildlife, through fish hatcheries, game farms, and scientific study.[47]

Such methods may have been foreign to the average sportsman in 1922, but the league's rhetoric — put forth in the pages of *Outdoor America,* a glossy membership magazine — was calculated to link these new tactics to old and familiar ideals. Under the spirit of "Waltonism," active management of fish and game would promote the sort of outdoor experiences that had defined the American character in the nineteenth century. Zane Grey supported the league's advocacy of "red-blooded pursuits" as an antidote to the "appalling degeneracy of modern days." Characteristic of Waltonism, Grey conceived of outdoor life as an exclusively male heritage: "If a million outdoor men who have sons, will think of these sons, and band together to influence other men who have sons — *then we may save something of America's outdoor joys for the boys.* "[48]

The league's founder and guiding light was Will H. Dilg, a Chicago advertising man and fervent bass fisherman. Fifty-three years old when he launched the group, Dilg was an excitable and forceful man, given to bouts of hyperbole both in person and on paper. He often said he had founded the group in memory of a son who had drowned, in order to perpetuate the spirit of boyhood in America.[49] A dozen of the league's fifty-four founding members worked in advertising or sales, and the Waltonians' tactics reflected this influence. Dilg imbued the group with a sense of small-town boosterism, beseeching middle-class businessmen to establish a chapter in every burg with a population of more than three hundred people. He recruited a stable of traveling speakers, including baseball commissioner Kenesaw Mountain Landis, and lured top writers — James Oliver Curwood among them — for the league's magazine. Under Dilg and his successors, the league was acutely publicity conscious, making use of newspapers and of the emerging medium of radio.[50] By 1925, the league had one hundred thousand members and had pushed its first major initiative through Congress: a law protecting a three-hundred-mile stretch of the upper Mississippi River and its backwaters, which were prime bass and waterfowl habitat. Dilg was exhausted by the legislative battle but apparently not surprised by the victory. He had, in 1923, declared that God had made the backwaters "for men like you and me" and that the Almighty stood with the Waltonians in preserving the area as a "playground."[51]

Taking its cues from natural-resource managers, the new league came out solidly for artificial propagation of fish and game. Fishing, for example, was more than a matter of natural felicity; in Waltonian parlance it was a challenge of engineering. In the face of the automobile and a swelling population, angling was bound to disappear unless the experts kept "ceaselessly at work on the sources of supply." To that end, the league enthusiastically supported the building of more federal and state fish hatcheries.[52] But the work of conservation could be undertaken by amateurs, too. Just after Christmas 1926, at the outset of a hard winter, the league appealed to its members to organize "feeding campaigns" for ground-dwelling game birds such as quail and pheasants. The root of the crisis was modern agricultural practice, specifically "clean farming," which had wiped out the hedgerows, weeds, and briar patches of old times. With farmers plowing to the edges of their fences, Waltonians had a duty to supply the winter food that the tractors had taken away. "Modernity has decreed through present agricultural methods that sportsmen must assist in feeding the game birds," the league declared in an urgent bulletin. Sportsmen in the new order were obligated under

Will Dilg (with fishing rod) and friends on a trip to the upper Mississippi River country. Photograph courtesy of Izaak Walton League of America; reprinted with permission of IWLA.

"DIVINE WALTONIAN PRINCIPLE" to mitigate the "harrowing effects" of nature.[53]

But the Waltonians' battle was bigger than propagation. Continuing a process begun in the nineteenth century, they sought to transform the value calculus of fish and game, away from the animals' subsistence value as meat and in favor of the sporting experience inherent in the taking

Local sportsmen plant trout in a stream in Waupaca County, Wisconsin, 1936.
Photograph courtesy of State Historical Society of Wisconsin, WHi (X3) 52197.

of wild creatures. Like the elite "gentlemen" sportsmen of days past, the
middle-class Waltonians looked down on the "game hog" who slaugh-
tered wildlife indiscriminately. And they were not shy about invoking
the police power of the state to enforce their views. In Wisconsin, for
example, Walton chapters that established private game refuges were
encouraged to have their members deputized under the local sheriff and
empowered to arrest trespassers.[54] While proclaiming their efforts as
purely public-spirited, the Waltonians clearly favored some publics over
others. About 1927, the national office distributed a political cartoon that,
unwittingly or not, dramatically illustrated the group's mission. It showed
a group of Waltonian men, respectable in appearance, standing with the
ghost of Izaak Walton at the edge of a forest. Arrayed against them was
a rabble of thuggish-looking characters whose labels included "Bad
Sportsmanship," "Poacher," and "Game Hog." Tellingly, the Waltonians
were excluding the undesirables by building a brick wall around the
woods.[55]

For all its vigor, Waltonian rhetoric underwent a marked transformation in the several years after the group's founding. Dilg and his fol-

lowers initially emphasized a Progressive "big stick" attitude, talking of fair play and good citizenship and damning corporate greed that was not coupled with wider ideals of the collective good. Dilg, for example, had in 1923 condemned the "nickel and dime chasers" who would drain the backwaters of the upper Mississippi, robbing sportsmen of their God-given paradise.[56] At a thousand-person Walton banquet in Chicago that same year, a clergyman had blasted the "traitors" of American business "who would for wicked gain blot out the stars if they could."[57] But the battle for the outdoors was not simply a matter of corporate greed versus clean American manhood. As Dilg and colleagues soon discovered, there was a natural — and highly lucrative — synergy between hunting and fishing and a dazzling array of products designed to make those pursuits more enjoyable. The average Waltonian might motor to the lake in his automobile, launch a mass-produced boat with a small outboard motor, then angle for bass with brand-name, nationally advertised tackle. Outdoor life, in short, was a business in itself. The league's magazine, *Outdoor America,* reflected this synergy. Without advertising, the publication could not survive. In 1925, with the Walton League suffering growing pains and a deficit, Dilg invited advertisers to display their wares at the league's convention. With rods and reels, guns, boats, and camping gear, the "Sportsmen's Show" would become a prime attraction of each year's meeting. Obsessed with boosting membership, Dilg asked Waltonians to remind their local businessmen that "the League is bringing tourists through their town who — spend MONEY."[58]

Awareness of business synergies was not confined to the Walton League. Much of Aldo Leopold's game research would be funded by the Sporting Arms and Ammunition Manufacturers' Institute. Cynical or not, this support reflected a simple truth: without something to shoot at, sportsmen would have no reason to purchase guns.[59]

The appropriation of outdoor experience for business use also could be indirect and "psychological," in keeping with the advertising trends of the 1920s. Advertisers during the decade learned to sell ordinary consumer products by linking them with human fantasies, fears, and desires. A car could convey sex appeal; a fine suit could keep a man's career on track; a cold cream might further a young woman's goal of marriage. Outdoor life — beyond its obvious selling points for makers of guns, tents, or fishing lures — might also be profitably associated with everyday goods and services. About 1923, the publishers of *Field and Stream* urged ad men to incorporate the "out-of-doors appeal" in their work:

> Out of this great love of man for Nature . . . what an opportunity for the advertising man to capture some of Nature's lore and turn it into advertising lure! — to turn a campfire by a stream into an advertisement for a slab of bacon or a pound of coffee, . . . to point majestic antlers against the purple of a twilight sky toward the sale of a motor car, a can of soup, a banking service or a cigarette.

Among stellar examples of such copy, the magazine pointed to an advertisement for Ivory Soap — a narrative about tired campers asleep "by a thousand starlit lakes and streams," the Procter & Gamble product having "bathed away their aches."[60] In the same vein, a 1924 ad for Whitman's Chocolates appealed to "that out-of-door craving for sweets"; it showed a fashionably dressed couple rather incongruously perched on a dock at a resort.[61] A 1926 ad encouraged vacationers to take along Camel cigarettes. Smokers leaving "civilization" and going to "the deep woods" were advised to pack several cartons.[62]

The trend was stretched to its logical — or illogical — extreme in December 1928, when the Winchester Repeating Arms Company launched its "Arctic Broadcasting" campaign. The gun maker arranged to sponsor fifteen nighttime radio broadcasts, allegedly beamed to the "hardy fur-clad pioneers" of the "Far North." Not coincidentally, sportsmen in more civilized locales were invited to eavesdrop on the messages being sent to Mounties, missionaries, and explorers. A print ad assured consumers that the "cabins, igloos, huts and ships" of these brave men were "well equipped with Winchester Rifles, Cartridges, Shotguns and Shells."[63] It was a roundabout way of selling weapons: invoking a mythical audience in a distant landscape that the vast majority of sportsmen would never see, all in the service of an outdoor pursuit — hunting — whose perpetuation depended on mythical, and much cherished, ideals of what it meant to be an American.

Wilderness Reborn

A casual sojourner might see "wilderness" in any stand of timber, but the acres that would come to be known as "wilderness areas" were something else entirely. Roadless and remote, they offered a primeval splendor that was lacking in the more obviously groomed landscapes of the regrown forest. The "wilderness," on first impression, might seem the polar opposite of planning, landscape design, or "artificialization" of resources. The word connotes a land apart, untouched, undeveloped, and unspoiled by human contact. To let alone is not to plan — or is it?

Beginning about 1919, when the idea of "creating" wilderness areas first gained currency, its advocates discovered that letting the land alone entailed a host of questions. These questions would be raised, not by the negative action of keeping hands off, but by an ambitious and sometimes heated effort to set the terms of wilderness preservation. Because the interested parties were widely dispersed, the key issues would be defined and debated mostly in the print media, both specialized and popular. What defined a "wilderness" anyway? For whose use would it be set aside, and why? Were the benefits simply recreational, or were they somehow psychic as well? Not surprisingly, the most effective wilderness advocates of the 1920s were not antimodernist dreamers, but foresters and landscape architects — people who knew how to shape and manage the land to provide human benefits. While the experience of traveling in the new wilderness might invoke pioneer days, the business of making this landscape was firmly rooted in the present. As its practitioners discovered, "wilderness management cannot be purely 'natural,' but must instead consist of a set of social and environmental compromises."[64]

On December 6, 1919, two foresters met to discuss the possibility that some areas within the national forests might deliberately be left undeveloped. Arthur H. Carhart, the younger of the two, was the newly appointed "recreation engineer" for the Forest Service in Denver, Colorado. (His colleagues, many of whom were old hands from the Pinchot days, had mockingly dubbed him the "beauty engineer.") The other man was Aldo Leopold, then a regional forest administrator in Albuquerque, New Mexico. A short time earlier, Carhart had advanced the idea — radical in its day — that the shoreline of Trappers Lake in Colorado's White River National Forest be kept free of roads or summer cottages. Leopold listened carefully as Carhart laid out his vision, then asked him to follow up in writing.[65]

Carhart's memorandum, written four days later, stands as a landmark document in conservation history. In it, Carhart conceived of "wilderness" mainly in terms of scenery. The Forest Service, he noted, had long managed its lands for timber, grazing, and flood control. But it had not, to date, delivered the "fullest return of scenic and aesthetic values" to the American public. Because many Americans desired to see land that was "untouched," the Forest Service might deliver this "fullest return" by letting some of its lands alone. This "natural scenic beauty should be available, not for the small group, but for the greatest population," Carhart wrote.[66]

Carhart's plan for wilderness must be read in context. Trained as a landscape architect, Carhart had been hired by the Forest Service mainly for reasons of bureaucratic jealousy. Stung by the increasing popularity of the national parks, the Forest Service in 1919 was scrambling to develop the recreational possibilities on its own lands, lest they be annexed by the Park Service. Carhart was striving to enunciate a policy of recreational use for an agency that had scarcely thought of the issue before. He was hardly in a position to advocate strict preservation, without regard to human utility. Focusing on human access, he was not averse to limited, rustic development that would help more people see the wild landscape. In 1922, for example, he set out his plan for what would become the major wilderness area of the Great Lakes region: Quetico-Superior, in northern Minnesota and Ontario. While keeping much of the area in a pristine state, the plan called for hotels, marked canoe routes, and "motorboat highways" by which tourists could be ferried into the backcountry.[67]

Leopold's definition of wilderness called for a more rugged landscape, less attuned to providing "scenery" than to replicating—as nearly as possible—the experience of traveling in a frontier environment. In 1921, he outlined his parameters for the place of wilderness in forest recreation. A wilderness area, he argued, should be "a continuous stretch of country preserved in its natural state, open to lawful hunting and fishing, big enough to absorb a two weeks' pack trip, and kept devoid of roads, artificial trails, cottages, or other works of man." Unlike Carhart, Leopold was not overtly concerned with issues of access. In fact, his wilderness policy was aimed specifically at the "substantial minority" of tourists who preferred to take their recreation without "automobile roads, summer hotels, graded trails, and other modern conveniences."[68]

Leopold recognized, better than anyone else in the 1920s, that creating a modern wilderness would be a matter of coining a historical, mythical, social, and economic rationale for the continued existence of wild land. Wilderness advocates, according to Leopold, had to invent a "mental language in which to discuss the matter."[69] In 1925 he published "Wilderness as a Form of Land Use" in Richard T. Ely's *Journal of Land and Public Utility Economics*. The choice of venue was ingenious. Beset by the tax-reversion crisis in the Great Lakes cutover, Ely and colleagues were struggling to define the terms of land utilization and planning and to account for social costs and benefits beyond the calculations of the marketplace. Leopold recognized that the value calculus for land use was changing, and he urged planners to account for factors that couldn't readily be quantified. Wilderness, he wrote, was "a distinctive environment which may, if rightly used, yield certain social values."[70]

Leopold's argument was bluntly historical, unabashedly male-gendered, and centered on human use. By implication if not by name, he invoked Frederick Jackson Turner. The encounter with wilderness had defined the American spirit. "For three centuries that environment has determined the character of our development; it may, in fact, be said that, coupled with the character of our racial stocks, it is the very stuff that America is made of. Shall we now exterminate this thing that made us American?"[71] Through canoeing, camping, and hiking, citizens could replicate the American encounter with the primeval, presumably reaping the benefits of psychological resilience and resourcefulness. Literally blazing their own trails, these new "pioneers" could avoid the increasing conformity of American social life. The modern wilderness would not be entirely undeveloped; for one thing, it would be protected from fire. (Elsewhere, Leopold even suggested that wilderness recreation and commercial logging might not be mutually exclusive, provided the logs were hauled out via water, instead of roads.)[72] Indeed, under enlightened management the wilderness could offer "an improvement on pioneering itself."[73]

All this, of course, made the modern wilderness different from the "real" one. Here, Leopold adopted the same stance he would with game management, which was to frankly acknowledge the tension inherent in collectively planning a landscape that was supposed to epitomize individualism. Any postfrontier wilderness was necessarily "artificialized," because it could exist only if shielded from the impulses of the free market, presumably through the constraining hand of government. When one forester, Manly Thompson, suggested that wilderness adventurers, of all people, should not need such "coddling and babying," Leopold's answer was forthright:

> Mr. Thompson wonders whether a self-respecting wilderness enthusiast would really want to play in a wilderness marked out and protected by a paternal government.
>
> It all depends on what is self-respect. Would a self-respecting athlete play at a game rather than wait for a real battle? Would a self-respecting boy fish in a wash-tub? Maybe, — if he has to. The capacity for illusion may not be self-respect; enthusiasm for half-loaves may be bogus; but the world continues habitable by reason of these failings.[74]

In the long run, Thompson lost and Leopold won — though the definition of "wilderness" and its purpose would be subject to constant reinterpretation and debate. Not until 1929 would the Forest Service headquarters in Washington issue standard criteria for selection and management of wilderness areas. More than six dozen would be designated by 1939.[75] In the 1930s, the wilderness concept would reflect a melding of romanticism with a zeal for primitive recreation, eloquently put forth by Robert Marshall and the Wilderness Society. In 1935, Leopold — pushing the boundaries of wilderness thought once again — went beyond the concept of human enjoyment and suggested that the wilderness should be preserved for its own sake, as an antidote to human "biotic arrogance." Because human beings could neither fully understand nor fully control the forces of nature, Leopold argued, wilderness areas should be allowed to stand as a token of humility and biological insurance.[76] The wilderness landscape existed as much in the human mind and heart as it did in physical reality; thus it could be remade while remaining untouched.

By the early 1930s, American outdoor life had been domesticated. What formerly had been private passions now were conducted under the guidance of the state, of commercial interests, and of a new class of managerial experts. Camping, hunting, and fishing were deemed too important to be left to the wishes of the individual. For the sake of democratic ideals, property rights, interests of order, and pecuniary gain, the rugged outdoors would never be quite as rugged again. Yet the new ordering of the landscape would help save vestiges of outdoor experience where there otherwise might have been none. Under the dream of continuous production, the landscape was managed to produce an annual crop of service to human beings, much as farm acreage produced corn or wheat. The task, as Aldo Leopold so acutely recognized, was to tame the forest while at the same time preserving at least the illusion of its wild past — to create a naturalistic landscape that constituted a new and improved human interpretation of nature itself. w

CHAPTER SEVEN

Harold Titus

"Old Warden" in the Woods

His book "Timber"...appeared at a psychological time....That book probably did more than any other single factor between 1920 and 1925 to give impetus to the conservation movement.

—BEN EAST, outdoor writer

Rube Pottle was a cutover prophet. For years he had been without honor in his own community, the former boomtown of Blueberry, Michigan. As a cutover land agent, he had peddled worthless acres to gullible settlers, only to see them fail dismally. This harvest of misery had converted Pottle to forestry; afterward he had campaigned tirelessly for fire protection and replanting of the pines. Scorned by his neighbors, he had endured their derision with patience and good humor. Always he likened his work to the painting of a picture. With foresight and planning, he insisted, the picture could be made real; it could rise up from its canvas and walk. Within a few years, his imaginary easel and paints had borne real fruit: the once forlorn Blueberry was producing a healthy annual "crop" of timber and tourists.[1]

Rube Pottle is even more extraordinary for the fact that he never existed. He was the hero of a short story. The prophet of Blueberry sprang full-blown from the imagination of Harold Titus, a Michigan writer and conservationist. He emerged from Titus's typewriter in 1922—a time when the magnitude of the cutover crisis was only starting to become clear. In the process of creating a fictional prophet, Titus had revealed the depth of his own thinking on the future of his beloved woods.

Harold Titus—author, conservation propagandist, policy maker, and journalist—possessed a singularly keen vision of the cutover's problems

Harold Titus. The forest, he believed, could be managed for sustained
yield of both material and spiritual riches. Photograph courtesy of Harold
Titus collection, Bentley Historical Library, University of Michigan.

and its promise. His prescription for the region, articulated first in fiction, then in magazine articles, was a delicately crafted synthesis. It combined the romantic fervor of James Oliver Curwood with the rationality of P. S. Lovejoy, without surrendering to the extremes of either. Like Curwood, Titus saw the forest as a deep reservoir of human spiritual renewal, even as a link to a romantic and mythical past. But he also shared Lovejoy's passion for methodology, recognizing that the cutover would have to be remade through coordinated human endeavor.

Titus was a conservationist for the New Era, seeking to preserve a modicum of individualism through the mechanisms of organization. As such, he displayed the planner's usual prejudices, particularly a belief in guidance by one's betters. But his vision was more expansive than crabbed. More than other writers on the subject, Titus knew that a single landscape could be invested with multiple meanings — indeed, that a multiplicity of meanings was crucial if cutover planners were to garner wide public support for their work. And he realized that even the most intangible products of the land — such as the satisfaction to be found in fishing — could be managed for maximum output. Going beyond Herbert Hoover's "production of leisure," Titus envisioned the efficient production of spiritual and mythical value in the forest. Thus he could claim, without irony or self-deception, that while his deepest motivations were "not economic," they also arose from the desire to make "efficient" use of cutover "waste land."[2]

Titus's two greatest contributions to conservation literature were written almost as afterthoughts, as diversions from a literary career that never really blossomed. The first was a polemical novel, *Timber*, published in 1922. The second was a series of articles for *Field and Stream* in the early 1930s, featuring a crusty and wise character known as the "Old Warden." In these works, Titus laid bare his desire for a centrally managed environment directed toward the production of agreed-upon social and economic ends. Titus's vision was delivered as light reading, but its core message was one of nuanced complexity, a blueprint for the cutover as a planned landscape and a strategy for making it real. His picture, too, eventually would stand up and walk.

A Writer's Education

Harold Titus was born in 1888 in Traverse City, Michigan, in the northwestern part of the Lower Peninsula. A rolling and verdant landscape, the

Grand Traverse area was (and is) a major fruit-growing region abutting the northern woods. As a boy, Titus witnessed the end of the lumber era. Later, he harbored vague but poignant memories of watching the region's last timber floating down the Boardman River toward the sawmills. His father died when Titus was a baby, but the boy still managed to get an education in outdoor life. As a youth, he was deeply influenced by the sporting magazines and their ongoing crusade against the "game hog," the indiscriminate hunter who killed without mind to conservation. Soon he was venturing afield with the local rod and gun club, helping to stock hatchery-raised trout in area streams. His boyhood adventures left a lasting imprint in two respects. They impressed upon him the idea of stewardship and care for natural resources; and they convinced him that the outdoor life was integral to all that was good and healthy in American manhood.[3]

At the same time, he was becoming a writer. As a student at the University of Michigan, he wrote campus news and sports for the *Detroit News*. When a bout with tuberculosis knocked him out of school, he went to work for the *News* as a reporter. Like many city-desk cubs, he functioned as a sort of male sob sister, covering sordid police stories and sensational trials. He witnessed the wrathful visit of Carry Nation to a Detroit saloon, writing up the incident with no small amount of humor. His two years at the *News* appear to have given him a fine sense of melodrama and the ability to write clean and quick prose.[4] As a hallmark of this training, he would become astoundingly prolific at writing formula fiction, sometimes churning out several thousand words in a day.

Whatever the satisfactions of newspapering, the country boy felt that the biggest stories lay beyond the city limits. Like many middle-class men of his generation, he sought adventure in the West. About 1912 he spent several weeks with a group of cowboys in Colorado, enduring bad food and January temperatures of forty degrees below zero. Though known to the cowhands as a "lit'rary guy," Titus endeavored to fit in, joining their boxing matches and winning a pair of spurs in a poker game.[5] The cowboy experience spawned three western novels, beginning with *I Conquered* in 1916. Titus married and settled in his native Traverse City, operating an orchard to supplement his writing income. By 1920 he was fairly well established as a writer of popular fiction, selling stories to major magazines such as the *Ladies' Home Journal* for as much as four hundred dollars apiece. Movie scenarios and screen rights to his books brought more income; his cowboy novel *The Last Straw,* for example, fetched $3,500 from the Fox studio in 1919.[6]

The cowboy tales were a means to an income, but Titus was grappling with bigger ideas. The young writer was flexing his intellectual muscles, trying to mold an ideology that was still in flux. At one point he met with the anarchist Emma Goldman. He submitted a story to Max Eastman's magazine, the *Masses.* He conferred with Sinclair Lewis. His intellectual streak vexed his editors, who scolded him for being too "psychological" and for favoring ideas over raw action. An editor at the Munsey Company admonished him to stick with stories about individuals, not groups or institutions. The literary world had one Upton Sinclair, the editor noted; it did not need another.[7] Titus was undeterred. As time went on, he did adhere more to the conventions of popular fiction, filling his tales with red-blooded men, virtuous women, and plenty of action. But he never gave up his advocacy of ideas or his long-term dream of literary greatness. In the meantime, if he could not be Upton Sinclair, certainly he could combine melodrama and message.

The world war, in which Titus volunteered and served stateside, gave shape to his amorphous thinking. By 1917, he was preoccupied with a number of questions: What was the connection between the individual and the state? What was the duty of one to the other? In an age of organization, how could the individual find fulfillment? What was the meaning of patriotism? He toyed with writing a novel — tentatively titled "The Dodger" — exploring the social impact of the war. Titus sought to transcend the "piffle of platform patriotism" and to plumb the real nexus between warfare and the individual. His hero would be a likable but noncommittal sort whose energies were awakened by the call to public service. The novel, Titus admitted, would be drawn partly from his own experience. He was wrestling with the question of what he owed his government and just how far the individual should surrender his identity to the imperatives of society. "Spiritual revolutions" were afoot, and the tension between duty and individualism was being stretched to its limits.[8]

As his writing made clear, this tension was equally applicable in peacetime. The Armistice did not make moot the question of what the individual owed society, and vice versa. Just after the war, Titus wrote a transparently autobiographical tale called "It Can't Be Done." Its protagonist is Jimmy Lawrence, a corporal just out of the army. His faithful dog at his side, Lawrence sets out on foot for the northern Michigan cutover. "[T]he big thing that impelled him was to seek refuge from people." During a rainstorm, he takes shelter with a farm family facing eviction. Lawrence reads the fine print in the family's lease and discovers that they have ninety days to raise the necessary cash. He confronts the landlord on their behalf, gets a reprieve, and vows to stay and help them win their fight. As he tells the grateful farm wife, the moral of the story is clear: " 'This dog and I . . . Why, we started out to walk away from other folks' troubles just as fast as we could,' he said to her; then to the dog: 'But it can't be done, buddy . . . it can't be done!' " [9]

Conservation posed similar questions of duty. Unlike most writers on the cutover problem, Titus actually lived in the forest region. Within a few miles of Traverse City lay a sickening abundance of fire-charred acreage. The advent of the Model T allowed sportsmen to go almost anywhere in this ruined landscape, and what they saw alarmed them. For Titus, as with most hunters and anglers, the immediate impulse was purely selfish; he was appalled at the destruction of grouse habitat and the degradation of the Boardman and other trout streams. But there were bigger motivations as well. Titus had been swept up in the talk of a "timber famine"; he also worried about the impoverishment of the cutover farming areas. Beyond the economic questions, he loved the forest for its mythical value—for its links to a wilderness past, for its part in forming the character of young men who hunted and fished as their ancestors had. Ironically, the preservation of individualism seemed to demand collective action. About 1920, Titus and friends formed a new "Conservation League" as successor to the old rod and gun club:

> There was only one thing to do, we said. That was to organize. . . . The outdoor magazines by that time had largely dropped their crusade against the game hog and were concentrating on more stringent laws, more hatcheries, more sacrifices on the part of sportsmen. . . .
> . . . It was a period when young hunters and fishermen sensed that an era had come to an end, that new conditions prevailed in the limitations of their outdoor fun and that no one knew just what those limitations were or how to adjust to them. [10]

Not content to rest on emotionalism, Titus began corresponding with experts in various facets of the cutover endeavor: forestry, fire protection, game management, and land economics. About 1920 he struck up a friendship with P. S. Lovejoy, who had just begun writing his path-breaking articles on land use for the *Country Gentleman*. Lovejoy's thinking lent a hard quantitative air to Titus's apprehensions about the cut-over. It also impressed him with the need to "translate" technical ideas for a popular audience.

In the meantime, Titus had been working on a story of the pinelands. He did not consider it his "big" novel in a literary sense. Rather, it was compelled by two pressing urgencies: the need for an income and the crisis in the cutover. The immediate motivation for its writing had been a trip to the Manistee River in 1918, when Titus was freshly returned from the army. There he had stumbled upon the ruin of a much loved forest tract; he had seen a whitetail doe swimming frantically down the spoiled stream, pursued by poachers' hounds. Like the young hero of "It Can't Be Done," Titus had realized that retreating from responsibility was no longer an option. To save the forest meant plunging head-long into the world of human affairs.[11]

In November 1921, *Everybody's Magazine* launched a five-part fiction serial about the cutover crisis. Titled *Foraker's Folly*, Titus's tale was part melodrama, part civics lesson. It was somewhat overwrought, rather bluntly didactic, and larded with implausible action. It was not great literature, but it was simultaneously instructive and entertaining. In 1922 it was issued as a book titled *Timber*. Its publisher called it "the first novel of conservation."[12]

The Lessons of *Timber*

The significance of Titus's fourth novel lies not in its artistic merit, but in its explication of a number of vivid metaphors concerning the cutover. These repeated themes and images presented the reader with an ideal-ized view of how the forest worked and how the landscape could be shaped to provide maximum human utility. Titus's most noteworthy contribution here was his all-inclusive sense of the human values to be found amid the pines. He was interested not just in board feet of timber, but in recreation, spiritual refreshment, and even the character-building experience to be found in forestry work. The forest, in short, could yield many products, both tangible and intangible. It was a veritable cornu-copia of the raw materials necessary to sustain a vibrant economy, well-balanced individuals, and a healthy body politic. This yield could be maximized through the agency of centralized, scientific management.

Timber tells the story of John Taylor, son of lumber baron Luke Taylor. At the beginning of the novel John Taylor is the dissolute son of a rich man, looking for an easy start in life. The elder Taylor gives his son three hundred thousand board feet of Michigan hardwood. Although this is much less than John had hoped for, he journeys to the cutover of Blueberry County in hopes of winning his father's admiration — and an inheritance.

John arrives in the cutover to find he has been duped. The local rail line has been pulled, with no hope of getting the hardwood to the mill. John is ready to quit when by chance he meets Helen Foraker. Helen is growing ten thousand acres of white pine as a crop — "Foraker's Folly," the planned forest begun by her late father. Helen helps John get his timber to market. Gradually she wins him over to her ideas of scientific, sustained forestry. John also sees that she has much finer mettle than his group of idle rich friends, epitomized by his grasping girlfriend Marcia Murray. They fall in love as the novel progresses.

The novel's chief villain is Jim Harris, lawyer for Chief Pontiac Power, head of Harris Development Company, and political boss. He wants to build a new courthouse, roads, and schools to entice unknowing settlers onto worthless cutover lands. The headquarters for his development company is a bucolic farmstead — but it looks good only because he has poured tons of expensive fertilizer into the sandy, acidic soil. Harris's scheme for public improvements will require a big levy on the county's remaining tax base, particularly Foraker's Folly.

Jim Harris and cronies, thwarted in their tax scheme, eventually resort to a sort of backwoods terrorism to drive Helen out. Meanwhile, John's father, the lumber baron Luke, arrives on the scene and contrives to get Helen's land for himself so he can cut the pine and relive his youth.

The climax of the book is a forest fire, set by backwoods simpleton Charley Stump, that destroys several hundred acres of Helen's land. Charley has acted at the direction of Jim Harris. The fire is put out by heroic effort, finally halted by a line of dynamite charges. Luke Taylor comes to realize Helen's merit and offers to bankroll her, though he still harbors some suspicion of her ideas, which he calls "moonshine."[13]

Timber is fairly well written and quite well plotted. Its structure is formulaic and sometimes clumsy, particularly in its heavy-handed use of foreshadowing. At times the characters speak past each other in monologues, carried away by their fervor for scientific forestry. Like many polemical tales, the novel suffers from a stark divergence between exposition and narrative. But overall it is an entertaining read, as Titus intended. He was driven not only by the imperatives of his pocketbook, but by the knowledge that the book's message would get across only if it was widely read. Titus considered it "a pretty good yarn" that "ought to interest people who don't give a damn whether trees grow anywhere or not."[14] Although its sales certainly did not approach the proportions of a Curwood novel, it had three printings by the end of 1922, remained in print until 1932, then was revived a year or so later because of booksellers' demand.[15]

Whether Titus's book was the "first novel of conservation" is subject to argument. Writers such as Stewart Edward White had raised forest themes as much as two decades earlier. While Titus was writing *Timber*, his fishing partner and fellow author James Beardsley Hendryx was making a conservationist out of the boy hero Connie Morgan. But Titus's book was unprecedented in its realistic treatment of the forest situation, specifically in the way it sketched a number of themes that would define the cutover effort in the 1920s and into the 1930s.

The Role of the Expert

Helen Foraker was raised in the woods, but she is hardly an isolated rustic. The shelves in her office are crammed with business books and forestry tomes. She knows that mere sentimentalism will not grow trees. She talks of "stumpage value" and "amortization." She quotes Swiss forestry experts from memory. All this knowledge, she says, resulted from an upbringing in which she and her father were regarded as oddballs by their Blueberry neighbors. Spurned by their community, they had looked to the wider world for enlightenment. As a child, Helen had played with "baby trees instead of dolls." The Foraker farm seems to have functioned as a sort of cutover Hull House, attracting international attention as a center of reform. "We never went anywhere to meet people; they came here: teachers of forestry, foresters from Europe."[16]

All this, of course, means that Helen is very different from the common people of Blueberry County, both in education and outlook. "She's stuck on herself an' won't mix with common folks," a Blueberry citizen tells John Taylor. This same citizen reports that Helen entertains a constant stream of "perfessers" from Ann Arbor—"Damn fools, all of 'em!"[17] The novel sets up a not-so-subtle class distinction between

Helen and almost everyone else in the community.[18] The cutover revival, Titus implies, must be a product of expert guidance; to thrive, the
community must submit to that guidance as a matter of faith. (Despite
the novel's realism on the subject, Helen Foraker does display two common but implausible attributes of fictional cutover heroes: she is independent of any bureaucratic apparatus, and she actually lives in the community she is striving to remake.)

The Obsolescence of the Pioneer

Reflecting Lovejoy's influence, Titus believed that the pioneer spirit had
outlived its usefulness. Haphazard land settlement led only to wasted
effort and broken dreams. This was particularly true in the cutover, where
much of the land was inhospitable to farming. In place of pioneering,
government had a duty to inventory lands and to guide settlers to acres
that were most likely to provide a decent return on investment.

In *Timber*, this idea is reflected in the person of Jenny Parker. Jenny
has come to the cutover with her husband, Thad, and together they have
purchased a plot of land from Jim Harris. But the land is unsuited to
cropping. Jenny and Thad are facing financial ruin. Their infant son has
died of malnutrition; Jenny has been told that she can have no more
children. (The symbolism practically boils over at this point: thanks to
the ruinous greed of Jim Harris, Jenny and the land are both barren.)
Helen Foraker and John Taylor visit the couple in their miserable shack.
Jenny, now dying herself, says she and her husband "believed we were
pioneers." But the cutover landscape has become "a graveyard for hopes,"
and the dream has proved to be a cruel deception. "That's what Jim Harris and all his kind are: murderers of hope!"[19]

The Dangerous Combination of Greed and Ignorance

In Titus's view, the occasional errant pioneer, though regrettable, was
hardly a social peril. The real danger lay in the co-opting of ignorant
people by the forces of large-scale capital.

Consider, for example, the rather transparently named Charley Stump.
At once both pathetic and contemptible, he spends his days pushing a
broken bicycle through the burned-over landscape. Symbolizing the absurd hopes of cutover agriculturalists, he is confident that prosperity is
just around the corner — meaning, in his case, that he'll be able to buy
tires for his bike. On his own, Stump is a harmless crank. At the hands

of Jim Harris, he becomes an arsonist and potentially even a killer. Similarly, the tragedy of cutover farming is compounded a thousandfold by big-scale land hustlers. Titus here clearly was taking another page from Lovejoy. Central to the cutover revival was the idea that capital had to be restrained, but also that unknowing individuals had to be protected from themselves.

The Fine Line between Paternalism and Contempt

The village of Pancake, Blueberry's county seat, is not a particularly appealing place. The streets are dusty, the buildings are run-down, and—most notably—everyone with any ambition seems to have fled. The sickness of the land is reflected in a malaise among its inhabitants.

Here Titus accurately foreshadowed a concern that would become acute by the early 1930s: without some infusion of economic stability, the cutover would degenerate into a socially malevolent landscape. The challenge for cutover planners, of course, was to prevent their well-meaning paternalism from becoming outright contempt for cutover residents. Lovejoy once fretted that the cutover would breed "paupers and morons and fires." Wisconsin's Richard T. Ely, though concerned for agriculture in the aggregate, made little attempt to disguise his scorn for cutover farmers. Titus himself never succumbed to the mean-spiritedness that was possible in cutover planning, but he did foretell the anxieties of rural sociologists (and eugenicists) before they became common currency.

The Forest as a Source of Moral Regeneration

For Titus, the cutover equation couldn't be explained merely with numbers. Some of it—much of it—was intangible. In one of *Timber's* most telling scenes, Luke Taylor is walking in a stand of Helen Foraker's replanted pine. Previously a stranger to scientific forestry, the old lumber baron discovers, if only fleetingly, that the trees he cut down in his youth can in fact be regrown for future generations:

> He looked up at the crowns above him, the whispering tops of the pine trees.... Something broke within him and light went from his eyes. Board feet! Always, he had looked at forests in the terms of board feet; today it was something else. There was more to this stand of baby pine than lumber, more than wealth.
> A breath caught in his throat and his eyes dimmed. He listened again and heard that time in the whispers of the tops an echo of his lost

youth. ... He moved toward the nearest tree and put out his hands as though for support. ... His palms pressed the bark on either side of the trunk; then stroked, gently, as a man will stroke some dear possession.

"Pine!" he muttered — "Michigan Pine! Oh, God — I thank you — thank you!"[20]

Helen Foraker recognizes this yearning, too, and she uses it to win Luke's financial help. The pines, she tells him, mean more than dollars and cents; they mean the history and lore of logging, they mean solace for the soul, and ultimately they mean "contentment."[21] The old logger opens his checkbook. John Taylor, too, is transformed by his encounter with the forest, from aimless playboy to vigorous woodsman and community leader.

Selling *Timber,* Selling Forestry

With the publication of *Timber,* Titus's love for the woods took precedence over his need for money, as it would for much of his career.[22] He had high hopes of "ringing the bell" with the novel, but by that he meant scoring propaganda points as well as sales. He had written the book in a sort of fever, he confessed to his publisher, racing to feed the public's rising interest in forestry. "I believe that next winter will see big things done and lots of noise made both by the real foresters and by the timber interests," he wrote in the summer of 1921.[23] At the time, talk of a "timber famine" had crested in Washington and in the states. Lawmakers were beginning to look beyond the question of depletion and toward wider issues such as land use, land acquisition, and forest propagation.[24] The United States had just entered a protracted agricultural depression, which was being felt hardest in marginal farm areas such as the cutover. Within a few years, the crisis in the Great Lakes states would become disturbingly clear. Already, the most prescient thinkers — particularly P. S. Lovejoy — were championing legal-economic remedies for the worsening situation in the former North Woods. The time was indeed right for *Timber.*

Titus launched his own promotional campaign, aimed at generating word-of-mouth notice among forestry professionals. He sent advance copies to the deans of the major forestry schools. Lovejoy, who had read the manuscript and served as technical adviser on the project, approved of the end result. He found the novel accurate and satisfying, "even if I do still flinch from some of the romance and movie situations."[25] With

Lovejoy's help, Titus approached Gifford Pinchot, who agreed to supply a promotional blurb for the book.[26]

The response from professional conservationists must have been gratifying. Readers of *Timber* included the officers of the Medicine Bow National Forest in Wyoming, who felt it treated the forest question "thoroughly and satisfactorily," according to their supervisor.[27] Ezra Levin, coordinator of Michigan's agricultural settlement program, thought Titus's work invaluable in putting "the land shark out of business. We enlist you as our chief aid right now."[28] John Doelle, Michigan's agriculture commissioner, read the novel in a single sitting and pronounced it "the *Uncle Tom's Cabin* of the conservation movement."[29]

The matter of impact is tougher to measure, but at the very least, *Timber* served as a backdrop for an intensified dialogue on forestry issues beginning in 1922. In June of that year, Lovejoy attended a forestry conference at Lansing, at which foresters and loggers reached consensus on the need for timberland tax reform and better fire protection. Lovejoy felt that Titus's work had resulted in "many a good crack at the inertia and bone-headedness of official Lansing. Nearly everybody at that June conference had read *Timber.* "[30]

At the beginning of 1923, *Timber* was released as a feature film titled *Hearts Aflame*. Titus had no involvement with the Louis B. Mayer production, other than receiving a much needed $3,500 for the screen rights. *Variety*'s reviewer praised the nine-reeler as "excellent melodrama with several splendid thrills."[31] Titus was not as enthusiastic. He found the movie inaccurate, and was appalled by a scene in which men toted boxes of dynamite through the flaming pines.[32] Yet the film would keep Titus's message in the public mind for several more months. Many years later, Lovejoy recalled that *Hearts Aflame* had been showing in Lansing when the Michigan Legislature made its first "decently generous" appropriation for battling forest fires.[33]

A New Vision for the Cutover

Today, *Timber* is a forgotten work, one of thousands of popular novels of the 1920s that litter the shelves of secondhand stores. What makes it special from the historian's point of view?

Consider the alternatives. P. S. Lovejoy sold reforestation based on hard economic merit, framing his arguments in ways that ordinary people would understand and find compelling. But he had a strong aversion to sentimentalism, believing that such "old maid" ideas only ob-

Anna Q. Nilsson as Helen Foraker (center) prepares to set off a dynamite charge to stop a forest fire in *Hearts Aflame,* the screen adaptation of Harold Titus's novel *Timber.* Photograph courtesy of Wisconsin Center for Film and Theater Research.

scured his message. James Oliver Curwood, on the other hand, saw the forest as a lode of mythical and historical riches, but he could mine those riches only by retreating into antimodernism. His exclusive focus on the heroic individual made him ineffective as a conservationist.

With *Timber,* Harold Titus bridged this considerable gap. Helen Foraker is a heroic manager, advocating modern methods to produce a steady stream of forest products — including timber, tourism, and the elusive but all-important "contentment." Once her vision is realized, her forest will yield not only raw material for the sawmill, but a balm for the anxious human soul. In *Timber,* Titus foretold the rise of the managed forest as an American social amenity, a space geared to consumption and pleasure as much as to efficient production.

At the same time, *Timber* recognized the inevitable trade-offs inherent in cutover management. The pioneer was dead; in his place rose a forest user whose actions would be guided by the omnipresent and (it was hoped) benevolent hand of the state. The planner's paternalism could easily give way to contempt.

The one aspect of reforestation that Titus did not see — or that he chose not to deal with in *Timber* — was the continued loss of local autonomy in a region that had long been at the mercy of outside decision makers. The forest's destiny would be determined by people who did not live there. Helen Foraker, the autonomous forester-citizen of Blueberry County, was not representative of the distant professionals who would remake the woods in real life. A novel about a bureaucrat would never have sold, of course. Like Lovejoy, Titus knew that the selling of the forest was an incremental job, and that unappealing realities sometimes had to be bent to meet narrative needs. With the writing of *Timber*, whatever its flaws, he had looked farther ahead than any other author. A decade later, he would advance the cause of natural-resource management again, this time in a national magazine aimed at sportsmen.

The Warden's Wisdom

About 1932, a folksy gentleman made his debut in *Field and Stream* magazine. He was a tall, lanky man, decidedly rustic-looking, with a battered old hat and an omnipresent pipe. Each month, in a fictional narrative based on fact, the "Old Warden" would deliver a tutorial on conservation.

His venue was not a classroom but a country store, where sportsmen gathered around a potbelly stove. His audience often included the town grouch, who complained about game shortages or intrusive rules that hindered hunting or fishing. But the warden handled each grievance with good cheer. Patiently, thoroughly, and in language the common man could understand, he set forth a rationale for modern management of fish and game. He advocated research into habitat and propagation, rather than just blind stocking of trout or birds. He spoke of the need for sportsmanship and self-control, stressing the mutual obligations of outdoorsmen as the nation became more crowded. Above all, he was an apologist for the "artificialization" of resources. To those who griped about the passing of frontier ways, the warden had a standing answer: natural resources were no longer free for the taking. To maximize each person's harvest — tangible and otherwise — nature's riches had to be actively managed. The pioneer legacy of hunting and fishing could be preserved only through the extension of state authority into the former wilderness.

The Old Warden (who appears to have had no other name) was the creation of Harold Titus, a frequent contributor to *Field and Stream* and later its conservation editor. Long experience in the outdoors, and on the Michigan Conservation Commission after 1927, had taught him that there were no easy answers to fish and game problems. The management of resources, Titus asserted, could not be left to armchair biologists. "The job belongs to the technical man," whose training could ferret out the "proper" answers to any conservation question.[34] The public had to acquiesce in this management, but it would do so only if persuaded to surrender some freedoms in the name of a common good. "I believe the problems and objectives of the outdoor technician need to be translated into the idiom of the average man who uses our fields and streams for the pursuit of—along with fish, birds and mammals—his immortal soul."[35]

As an ambassador of modernity, the Old Warden was deliberately and cleverly crafted. He was homespun and home-schooled, gaining his conservation knowledge through reading in the evenings. Taking a page from Lovejoy, Titus knew that his fictional lecturer could not be an off-putting college man. Instead, he was a smart but practical soul who had looked at the new conservation and decided that it made sense. The boys around the potbelly stove—and in the *Field and Stream* audience—were more likely to heed the words of a man who confessed that, when reading technical reports, "I have to go at 'em with a dictionary on one knee."[36]

Aldo Leopold, the foremost thinker in game management circa 1930, often was too detached and scholarly to appeal to the average sportsman. But as "translated" by the Old Warden, his research took on a friendly tone. Explaining Leopold's work on prairie chickens, the warden made a case for state-funded habitat management and artificial feeding stations. The birds, he said,

"will fly five miles to get to the dinner the state offers 'em....
"Who knows but, a few years from now, we'll have so much cooperation among these states that a chicken can get up and go where he pleases and never be out of flying distance of food and shelter? Be a grand thing, wouldn't it boys?"

This potential bonanza had its cost, including bag limits and license fees. For the town grouch, who pined for the days when "everything was free," the warden had a quick rebuttal: "There's too much lookin' ahead to be done in this conservation job to have much time for lookin' back."[37]

In speaking through a fictional warden, Titus revealed much about his own role in conservation. Game wardens, like forest rangers, were the advance guard of law, science, and administration in what formerly had been an unmanaged landscape. Their guise of rugged independence was mostly illusory, but it allowed them to play a conciliatory role in accustoming the public to changes that were unsettling to many. The warden supplied a human face for what otherwise would have been a distant and unfathomable bureaucracy. Though his message was sometimes unwelcome, it also could be delivered in ways that ordinary people might come to understand, even accept. The warden's job—and the writer's—was to advance the imperatives of modern management, while persuading the public that the trade-offs they entailed were worth it.

CHAPTER EIGHT

Spiritual Means to Economic Ends

Defining the Forest Fire

> Suppose that . . . there should be created tomorrow a body of 50,000
> forest landowners. . . . Do you realize that as long as we have 50,000
> forest fires a year, there is, for every one of these owners to engage in
> forestry, another man with a torch waiting every year to destroy his
> enterprise?
>
> —E. T. ALLEN, "50,000 Firebrands," c. 1926

> [T]he destruction of the wild things of the forest by the red
> demon . . . coming into the green canopies and sanctuaries . . . is
> brought in largely by man, who boasts of a far higher intelligence
> than the creatures of the forest.
>
> —JOHN D. GUTHRIE, "Fire in the Sanctuary," 1928

Random, widespread, uncontrolled fire is an enemy of rational forestry.
Its most serious material effect is not the destruction of standing timber, but the discouragement of future investment in tree crops. Without a reasonable degree of protection from fire, no one — in either the
private or public sector — will find much incentive to plant a crop that
may take fifty or a hundred years to ripen. Forest-fire prevention and
suppression, whose modern methodology was conceived largely in the
1920s, is a matter of organized risk abatement, with the aim of encouraging capital investment in the workings of nature.

Bernhard Fernow, a pioneer of American forestry, had fumed in 1890
that "the whole fire question in the United States is one of bad habits
and loose morals."[1] A generation later, Fernow's colleagues had not lost

their sense of indignation. But they also were targeting the fire problem with the technocratic mechanisms of law, science, and administration. Stopping fires, much like growing trees, was a question of bringing collective will and intelligence to bear in a formerly uncoordinated landscape. Fire, "long regarded like tornado or flood as an act of God," by 1920 was known to be "due to human negligence and incompetence," P. S. Lovejoy wrote with certainty in the *Country Gentleman*. Once it was targeted by forestry's managerial elite, the fire scourge was "preventable and curable."[2]

Lovejoy's confidence was not entirely unfounded. By 1930, state, federal, and private experts had built the foundation of an effective fire-control system. Its cornerstone was prevention, in the form of a concerted mass-media campaign to alert the public to the danger of fire. When fires did start, they were detected and doused by a newly aggressive suppression mechanism. With cooperative state-federal funding for lookout towers, forest roadways, and manpower, the effort managed to reduce the size of the average fire substantially. In Michigan, for example, forest fires burned nearly three-quarters of a million acres in 1925, the peak for the post–World War I period. By 1929 the acreage burned had fallen to 290,000, even though the number of actual fires was higher.[3] Nationwide, the fire menace was more intractable, particularly in the West and the South. But here, too, an aggressive prevention and suppression campaign would begin to show results after the early 1930s.[4]

By 1935, forester-writers had begun to assemble the first rough draft of the struggle's history. Much like reforestation, the fire battle was posited as a technical problem in need of expert solutions. In *American Forests* magazine, Ovid Butler wrote proudly of the American "fire-fighting army" that in peak season numbered twenty thousand men. The capital investment in towers, roads, and other fire-fighting necessities was $125 million, all aimed at the defeat of "forest enemy number one."[5] Stewart H. Holbrook, a lumberjack who became a professional writer and editor after 1923, made considerable hay of the forest fire topic, eventually writing the first book-length history of wildland fire in America. In *Burning an Empire* (1943), he lauded the "greatest forest protective force on earth" and attributed the persistence of fire to "man's carelessness and superstition and plain vindictiveness."[6]

The real story is not so simple. Since colonial times, forest fires, much like the forest itself, have been a culturally constructed phenomenon.

Although fire itself is a process of nature, the exact terms of its presence in America have mutated over time, in response to changing economic factors, methods of logging and other forest use, and agricultural practices. (And although fire may rage independent of man, the setting of fires is another matter: during the 1920s, the majority of forest fires in the Great Lakes states were "anthropogenic," or caused by people.) Further, it is hardly correct to imply, as Ovid Butler did in 1935, that fire has always been universally regarded as an "enemy." Before the twentieth century, when the forest was seen largely as an obstacle to progress, fire was viewed with alarm only when it threatened property or human life. Fire, as the Indians knew well, could actually be a useful tool for clearing land, driving game, or altering animal habitat. After 1900, fire prevention and suppression became handmaidens of industrial forestry, which sought to capitalize natural forces for the long-term growing of tree crops. In the 1920s and later, this theme would be echoed by tourism interests, which were awakening to the value of fire control in encouraging the capitalization of the woods for recreational ventures. But public attitudes, as P. S. Lovejoy discovered during his tour of "Cloverland" in 1919, were slow to change. A lot of people in that smoke-ridden paradise believed that fires were "at least harmless and often highly desirable in that they make the land easier to clear."[7] The high-pitched publicity campaign for forest-fire prevention, launched by groups such as the American Forestry Association in the 1920s, was not— as Ovid Butler would have liked to think—an appeal to universal, timeless ideals of conservation. It was, rather, an intensive reworking of the fire idea in America, with aspects that were diametrically opposed to frontier practices and attitudes.

The story of fire parallels that of forestry, in three stages. First, the economic and material factors of frontier land use sparked a series of spectacular "holocaust" fires in the Great Lakes states beginning in 1871. Second, the emerging forestry profession made fire control a key element of its organizational culture after about 1900, recognizing that the forest could be managed for maximum utility only if it was kept from burning. Finally, in the 1920s, professional and popular rhetoric on the fire problem diverged sharply. The rubric of risk abatement, much like talk of a "timber famine," was too abstract to capture the attention of the broad American public. Instead, the antifire press campaign (as orchestrated by foresters themselves) would frame fire as an "evil" that threatened to destroy a romantic and sentimental image of the woods. As in the larger forestry crusade, the fire campaign would invoke nostalgic, wistful, even irrational imagery with the aim of engineering a closely managed, predictable, and continuously productive natural environment.

The Economic Culture of Fire: From Peshtigo to Cloquet

On Sunday evening, October 8, 1871, a "Great Tornado of Fire" roared into Peshtigo, Wisconsin, a lumbering town about fifty miles north of Green Bay. The effects were so horrible as to be almost beyond comprehension. As the conflagration tore through the streets, dozens of people took shelter in a lumber company's boarding house and were incinerated in an instant. Hundreds more jumped into the Peshtigo River. Some survived, but many more simply suffocated as the blaze sucked the oxygen from the atmosphere. Others were swept downstream and over a lumber mill's dam, which itself was on fire. Within an hour, the town was gone. In the rural districts, entire families were found dead, some having committed suicide rather than submit to the flames. The Peshtigo Fire killed more than a thousand people and burned more than a million acres of forest. It was the worst tragedy by fire in the history of the United States. Some of the survivors had been sure that the flames were those of Armageddon and that the entire world was being consumed.[8]

One might think that the Peshtigo disaster would have spawned cries for forest-fire protection, much as the Chicago Fire of the same night led to demands for fireproof building construction. Yet it was not so. Why?

The answer was not, as Ovid Butler and colleagues liked to believe, that fire was simply an outgrowth of "wasteful" or "neglectful" forest practices that later generations would wisely prohibit. Fire, at the time of the Peshtigo tragedy, was conditioned by the unique circumstances of frontier land use — circumstances that would disappear only as the frontier did.

Among the nineteenth-century factors were these: First, Great Lakes timber at the time of the Peshtigo blaze was so plentiful as to constitute a free good, at least for the lumbering interests that had sewed up large tracts of forest land. To use the foresters' parlance, trees were "mined" rather than grown as a crop. Fire protection, which made economic sense only as a long-term form of risk abatement, held little appeal in this atmosphere of "cut and run." Second, the chronic American scarcity of

Forest-fire damage at Carlton, Minnesota, undated. Photograph courtesy of State Historical Society of Wisconsin, WHi (W6) 12674.

capital and labor also encouraged quick exploitation of the forest. The choicest trees, especially white pines, were skimmed from the land first, with lumbermen leaving a residue of "slash," or cut branches, behind as waste. In some spots, the forest floor might be piled twelve to fifteen feet high with flammable debris. Third, fire itself was viewed with considerable ambiguity by forest pioneers, who were struggling to subdue the woods and establish farms. Though it sometimes threatened life and property, it also was a quick, if crude, method of clearing land. And given that the dominant social paradigm, as late as 1925, held that the woods must give way to agriculture, the burning of the forest was not often seen as regrettable. Awful as were the holocausts at Peshtigo and elsewhere, if the forests were fated to vanish, then forest fires were merely a temporary phenomenon that must accompany the conquering of the continent. One man remembered the situation in the Michigan woods during the 1890s: "My father used to become a little concerned about a fire getting too close into our town. I remember he and I were walking around one time and we found a man with a shovel who was putting dirt on a fire. That was enough of an oddity for me to remember—the fires just burned unless they got into a town."[9]

The prescient few who raised the fire question after Peshtigo tended to view it in moral and educational terms, as if sufficient publicity and law enforcement would snuff out practices that ran counter to plain sense. "There is no *necessity* for this *criminal* negligence," a Wisconsin forester declared in 1882. While advocating regulation of forest practices, such as slash disposal, this man saw the problem mostly as one of awakening citizens to obvious truths. "Not one person in ten has any idea of the necessity of care as to forest-fires, and it all comes from ignorance."[10]

In actuality, not only did the social menace of fire have to be consciously defined, but the mechanism of suppressing fires once they started had to be built from scratch, as well. Here, the challenges facing nineteenth-century foresters were insurmountable. The law had been targeted at fire protection (with varying degrees of enthusiasm) since colonial days, but effective fire control also needed the application of science and administration. Forestry as a professional discipline was non-existent in the United States before about 1900, except as a minor adjunct of horticultural science. And state and federal governments were sadly lacking in the two prerequisites for a modern administrative state: a broad tax base and a specialized managerial workforce. Perhaps it was no wonder that a man of Bernhard Fernow's erudition would fulminate about fire being a problem of "loose morals": in 1890, it really was the only avenue of attack he had available.

In the meantime, the holocaust fires of the Great Lakes states continued unabated. When a firestorm engulfed Hinckley, Minnesota, in 1894, it was met by some two hundred volunteers of the Hinckley Fire Department, who of course were no equal to the flames. The blaze killed 413 people by official count; the actual toll may have been as high as 600, because entire families in the back woods were burned to ashes and never identified. The Hinckley blaze was the first to receive widespread national press coverage. Unlike the remote settlement of Peshtigo, where fire had struck at night, Hinckley erupted during the day, and there were plenty of survivors to tell harrowing stories to the reporters who rode the trains up from Saint Paul, just eighty miles away. One effect of the coverage was the appointment of Minnesota's first state fire warden, Christopher C. Andrews, who was granted a budget of six thousand dollars. Andrews later surmised — probably correctly — that in the absence of adequate money or manpower, his biggest contribution was in publicity, "to bring the cause of forestry in all its phases before the public."[11]

The last of the Great Lakes holocaust fires, defined as those that claimed hundreds of lives and burned hundreds of thousands of contiguous acres, occurred in northeastern Minnesota in 1918. It leveled the lumber town of Cloquet and invaded the outskirts of Duluth, sending thousands of refugees pouring into the neighboring city of Superior, Wisconsin. Several hundred people perished; again, as at Hinckley, the exact amount of backwoods carnage was beyond reckoning. One writer blamed the tragedy on a "pennywise legislature," which had cut the forest ranger force in half.[12]

Yet the likes of the 1918 fire never happened again, and the reasons lay mainly in shifting economic realities. The region's virgin softwood was almost entirely gone by 1910. The cut-and-run practices of the frontier had given way to more intensive logging, including the harvest of former "weed" trees for wood pulp and fibers. Cloquet's lumber interests, seeking a measure of permanence for their community, rebuilt it on a foundation of scientific forestry. Specifically, they pioneered in the use of wood chips from aspen, jack pine, and balsam for new "synthetic" lumber — the forerunner of today's composite pressboard. The Wood Conversion Company, as Cloquet's immense new factory was dubbed, began turning out products in 1922. As opposed to the rapacious lumberjacks of Paul Bunyan lore, the citizens of Cloquet increasingly saw their future as "scientific users of wood." Not surprisingly, the same timber companies that had shown little interest in fire protection just a few years before now became active supporters. Once their enterprise was conceived of as permanent, the challenge of risk abatement suddenly looked like good business.[13]

Defined by Fire

Stephen J. Pyne, the foremost historian of American wildland fire, sees a fundamental turning point in fire history in the early twentieth century. Americans had long struggled to subdue the land and convert it to agriculture, through individual effort encouraged by public policy measures, such as the Homestead Act of 1862. But sometime after 1900, at different times in different places, an imbalance in American land use became apparent. Experts believed there was too much land in crops, not too little, and that less fertile acres were returning too little money to repay their owners' sweat and capital investment. At the same time, the products of the forest — from timber to flood control — were becoming increasingly dear. The result was a concerted effort, through government and private expertise, to remove "submarginal" agricultural acreage from food production and devote it to forest crops. (This phenomenon also occurred spontaneously through the working of economic forces. It was seen, for example, in the proliferation of abandoned farms in New England about 1900.) Displaying a certain conceptual elegance, Pyne refers to this historic shift as the "counterreclamation."

The counterreclamation posed a challenge to the traditional conception of wildland fire in the United States. A fire in the woods, formerly regarded as an agent of agricultural manifest destiny, now was defined as an enemy of permanent, predictable forest production. The rural boosters and backcountry farmers who had regarded fire as tolerable, even useful, soon were exhorted to embrace a policy of total fire exclusion. Far from being universally accepted, the modern fire ethos would trigger a severe cultural clash that, in some places, would linger into the New Deal era.

Fire — dramatic, visceral, and readily quantifiable in terms of dollars-and-cents damage — became the rallying point for the new profession of forestry. Fire control constituted the "greatest single benefit" to the public in forestry operations, according to the handbook issued to federal forest rangers in 1906.[14] Timber interests were indifferent or even hostile to foresters initially, but the more enlightened loggers eventually saw a connection between sustained-yield forestry and economic self-interest. Without this conception of economic permanence, Pyne asserts, the forestry profession could not have existed:

> [P]rofessional forestry found that its greatest ally was industrial logging and its greatest enemy the frontier. So long as logging was considered as a form of frontier landclearing, there was little need for foresters. Only when it became linked to an industrial economy dependent on raw forest materials — a condition that brought the need for continued or sustained yield production and for certain standards of efficiency — did logging cease to migrate. It became instead part of the counterreclamation.[15]

The fire issue also played well from a public relations standpoint. A few rural renegades may have insisted on torching the woods, but for foresters after about 1910 a bigger concern lay in defining the forest fire for an increasingly urban population. E. T. Allen, who headed a consortium of private foresters in the West, emerged as a leading antifire propagandist under the guidance of George S. Long, general manager of the Weyerhaeuser Timber Company. Fire protection was "the strongest game" available to publicists who were selling forestry programs to the public, Allen believed; it was "easiest understood as well as desirable." His organization's aggressive publicity apparatus included a "general news bureau" to turn out press releases. Fortuitously or not, a series of disastrous fires in the West in 1910 helped to focus Allen's message and to win support for the Weeks Law of 1911, which established the framework for federal-state cooperation in fire control.[16]

Still, fire programs could be a tough sell, especially in places where the counterreclamation did not take hold until the 1920s. One such place was the Great Lakes cutover. C. L. Harrington, Wisconsin's chief forester, found himself having to dance delicately between the demands of industrial foresters and of cutover settlers who relied on fire for land clearing. In 1921, many cutover farmers believed that fires were "a blessing, and in many instances pay no attention to them," Harrington wrote with evident frustration.[17] Like his friend and colleague P. S. Lovejoy, Harrington felt compelled to pay obeisance to frontier practices while waiting for the counterreclamation to kick in. "We all want to see settlers come into the state and land cleared, but there is no sense in burning up a lot of country" in the process, Harrington advised in 1922.[18] Even for Wisconsinites who were not cutover pioneers, fire might seem relatively benign as long as the forest was expected to yield to the plow. Anita Bowden, a young woman who made an automobile journey through northwestern Wisconsin in the bad fire year of 1925, recounted seeing forest and grass fires all along the route. Driving toward Superior on a deserted road after nightfall, she noted the red glow of countless cutover blazes in the distance. To a modern reader, her account of this surreal landscape seems oddly nonchalant. "Someday it will be a wonderful farming section, I suppose."[19]

For foresters, the biggest challenge lay in the American South. There, it was common for upland farmers to torch the woods periodically to clear the underbrush for free-range grazing. The practice—which eventually would be shown to have certain ecological benefits—was a threat to industrial forestry, as fire sometimes got away from farmers and into adjacent commercial timberlands. Because many rural southerners were isolated or illiterate, an antifire message in newspapers and magazines could not reach them. The U.S. Forest Service (USFS) sent a traveling lecturer, H. N. Wheeler, into the region as early as 1924. Wheeler was an evangelical Christian full of "Billy Sunday emotionalism," as one colleague put it. To the considerable discomfort of some foresters, he skillfully incorporated the woods-burning topic into a larger theme of incendiarism, namely the type wrought by hellfire.[20]

The southern problem also raised an alarm at the American Forestry Association (AFA), which was closely allied with industrial forestry interests. Ovid Butler recognized that the fire message, in any medium aimed at the general public, had to be clothed with elements of entertainment, even melodrama. In 1928, under Butler's supervision, the AFA launched its Southern Forestry Educational Project, the most extensive and innovative publicity campaign it had ever undertaken.

The cornerstone of the AFA's southern crusade was film — specifically, dramatic and comical films shown to rural schoolchildren and adults. Butler and colleagues initially relied on educational films from the Forest Service, but found them didactic and dull. So in 1929, the foresters went into the moviemaking business themselves. W. C. McCormick, regional director of the project, recruited a cast and crew among foresters in Georgia and Mississippi. On June 7, 1929, the amateur troupe began shooting *Pardners*, a melodrama directed by Erle Kauffman, a writer for *American Forests and Forest Life*. The resulting film, the first of several AFA cinematic efforts, was fast paced and anything but subtle. It featured constant fires, numerous explosions, an eight-year-old girl trapped in a burning shed full of dynamite, and an epic hero-villain struggle — a "tree-growing, fire-hating, and peace-loving farmer" versus a "woods-burning murderer."[21] Focusing on Georgia, Florida, and Mississippi, the AFA sent several teams into the South in specially equipped film trucks. The movies were supplemented by lecturers, placards, pamphlets, and press releases, with the antifire message eventually reaching more than two million people. "Certainly 'Pardners' is going over in great shape with every type of audience," Ovid Butler reported after an inspection tour in the fall of 1929.[22]

It is difficult to gauge the effectiveness of the AFA's three-year southern campaign. The frontier custom of "firing the woods" was too deeply ingrained to be wiped out quickly; indeed, Butler and others recognized that the main targets of their crusade would have to be children, whose minds were still malleable. At times during the late 1920s, the Southeast would account for as much as nine-tenths of the forest acreage burned in the United States, and a statistic of that magnitude was not going to be eradicated overnight.[23] Sometimes it was hard for the foresters to conceal their contempt for "inbred" southern habits that ran counter to the professional ethos of fire exclusion.[24] In time, when research showed that controlled burning could be beneficial for forest regeneration and game management, it would be undertaken by experts. Woods-burning by unlettered individuals would become an anachronism, along with the idea of local autonomy in the management (however crude) of natural resources.

Professional Practice and Popular Imagination

The emerging practice of fire control had several elements. The most central of these was information. A grim fact of forest-fire science was that small blazes could grow big at an exponential pace. Much of the physical development of the fire-fighting mechanism was directed toward spotting fires early on and relaying information that would be used in bringing men and machinery to bear against fires before they could spread. Wisconsin, for example, built its first fire lookout towers in 1912 and experimented with airplane lookout patrols as early as 1915.[25] Access to the woods also was crucial; fires could not be reached by sufficient forces unless roads or trails crisscrossed each forest tract. (An alternate solution to this problem would be the development of the parachuting "smokejumper" in the 1930s.) Administrative remedies had to be divined for manpower issues — namely, the challenge of ensuring a large and reliable per diem workforce that might be called on just a few days a year. The fire-suppression apparatus, Michigan forester Russell Watson noted, "must be able to almost instantly expand to meet any situation; to expand as fire does, i.e., with geometric rapidity."[26] Finally, the basic science of fire had to be coupled with techniques of management to determine when and where the fire risk was highest and to deploy fire-fighting resources accordingly. Fire control, Watson asserted in 1922, was not a "sentimental" affair but a "business proposition," and without it, no one would dare invest in the "big capital" that a crop of trees represented.[27]

The catch, of course, was that fire prevention — and the political consensus needed for funding of public fire-control mechanisms — lay not with a few industrial foresters but with the general public. This necessitated a retuning of the fire message, away from hard economics and to-

A local protection committee fights a fire at Laona, Wisconsin, 1933. Photograph courtesy of University of Wisconsin–Madison Archives.

ward the spiritual meaning of the forest. As foresters solidified a policy of fire exclusion on behalf of industrial forestry in the 1920s, they simultaneously sought to demonize fire in the public mind. Fire, in this context, was portrayed as a threat to the beauty and permanence of the forest.

Foresters and their allies grappled for novel methods of getting the fire message across. A Pennsylvania forester, Chas. R. Meek, suggested that antifire slogans be spread throughout the woods on signs and posters, where such "propaganda" would resonate directly with hunters and nature lovers. Pennsylvania already had sent a Reo Speed Wagon into rural communities to show antifire movies, and the "new stunt" of radio broadcasting held considerable promise, Meek reported in 1923.[28] Some authors found it handy to imbue the work of fire fighting with elements of romance and individualism. The rangers who battled blazes in the northern Minnesota "wilderness" were among "the last of the real woodsmen," a magazine writer reported; they shared a kinship with the rugged men of the Old West, except that their horses had become canoes.[29]

Children, who constituted a captive audience in the schools, frequently were targeted with antifire messages. In 1922, USFS chief William B. Greeley described the nation's schoolchildren as an "army of 22,000,000," which must be enlisted to "fight a national foe that ravages the land before our eyes."[30] (He was not the first, nor would he be the last, to link the fire crusade with good citizenship.) The American Forestry Association published a mystical little "masque" for production as a school play, featuring anthropomorphic trees that trembled at the mere mention of "[o]ur enemy, the Fire." As described by the playwright, Fire itself was a role that every fifth-grade prankster must have coveted: "Fire wears grotesque sack-like garment of red cambric with red or yellow pasteboard fins attached to back. Hands and feet should be free for crawling."[31]

The Boy Scouts, whose ethic combined a love of nature with a quasi-military emphasis on citizenship and physical vigor, were a natural recruiting ground for antifire crusaders. Gifford Pinchot, while serving as Pennsylvania's chief forester, had distributed one hundred "gold medals" to individual Scouts for meritorious service to forests each year in the early 1920s. Pennsylvania Scouts also were enlisted as "forest guides," with pins and decorations being issued to those who signed pledges and performed forest-protection tasks.[32] In popular literature for boys, fire fighting was posited as a grand adventure with overtones of duty, service, and bravery. Scoutmaster Robert Shaler, who wrote a series of novels featuring Boy Scouts as small-town heroes, in 1915 had sent his fictional troop marching off to do battle with the forest enemy:

> The cry of Paul Revere in those old Continental days could hardly have thrilled . . . more than the brazen notes of that [fire] bell did the gathering scouts. . . .
> [I]t was hardly to be expected that any mother would dream of objecting to her boy going to the front no matter how her fond heart might be gripped with natural fears. Pride would step in and make sure that the finger of scorn should never be pointed at *her* boy.[33]

For children and adults alike, a key selling point emerged in fire's threat to woodland animals. The tactic was doubly useful because it could be applied with equal utility to sportsmen and to nonhunting nature lovers. Forester John D. Guthrie was among the most skillful of antifire propagandists on this theme. In a piece for the Izaak Walton League's magazine in 1928, he eulogized the countless forest denizens that "had been caught, trapped, and crisped, and had paid tribute to the fire god." The piece was augmented by photographs of fuzzy foxes, a spindly fawn with demure eyes, and two bear cubs, "[l]ittle playboys of the forest wilds."[34] Never mind that game management in the 1920s was just barely a science, much less one that could accurately assess the impact of fire on wild animals. Even Aldo Leopold, who seldom let his passions overrun his sense of scientific detachment, weighed in on the topic for Ovid Butler's *American Forestry*. Leopold took a cautious approach, stringing together anecdotal evidence from rangers and concluding—tentatively—that fire could do substantial damage to game stocks. Butler displayed no such reticence. He illustrated Leopold's article with stark and disturbing photos. A picture of a fire-charred Canada goose, which apparently had perished because she refused to leave her eggs, was cited as "tragic evidence of how the red scourge of field and forest is destroying our wild life."[35] To borrow E. T. Allen's phrase, the animal angle seemed perfectly calculated "to get good, warm human emotion aroused to its proper place in viewing the forest-fire picture."[36]

Fire also was a ready-made, enthralling climax for any work of forest fiction or cinema. Harold Titus's novel *Timber* had featured a ferocious blaze, though apparently not ferocious enough for Louis B. Mayer, who jazzed it up for the movies by arranging for a locomotive to drive through the flames. One writer noted in 1926 that the fire scene had become almost de rigueur in the forest movie genre. Hollywood, of course, tended to make fire fighting look more glamorous than gritty.[37] For the film production of James Oliver Curwood's *Nomads of the North* in 1920, crews erected a phony forest at Universal City, California, with fake trees that had to be "trimmed with natural foliage, planted in the ground, barked and painted." Before torching the ersatz woodland in the picture's climactic scene (while six cameras whirred), the filmmakers had to dig a firebreak all around it, lest the blaze spread to real timber nearby.[38] The fire scene became such an American convention that, when it did not exist, it had to be manufactured. Felix Salten's novel *Bambi*, imported from Austria and translated into English in 1929, had no such scene. The blaze in the cinematic *Bambi*—now permanently seared into the imagination of three generations of moviegoers—would be fabricated by Disney animators in 1942.[39]

A Change of "Mental Habits"

The antifire message did not immediately take hold among the public in the 1920s, but if nothing else it did crystallize the realization among foresters that public relations was a vital component of their profession. Wallace Hutchinson, who was one of just a few trained foresters to practice public relations full-time, wrote in 1931 that his colleagues were beginning to recognize that Americans would not "worship at the shrine of forestry" unless they were given a compelling reason. The forester "must preach the gospel of forestry from the housetops that all may hear and understand," Hutchinson asserted.[40]

Especially in fire control, the expert could not succeed without enlisting the general public. If forestry was conceived of as an arcane science, Erle Kauffman wrote in 1928, the public might not realize that "individual carelessness" was the single greatest cause of fire. Instead, citizens had to be persuaded to "fight side by side" with foresters. "The time will come when we shall be a nation of foresters, and when the constructive administration of our forest land will be formulated into a romantic and prosperous pastime."[41] Platitudes and exhortations were not enough, one forester noted: "Our real educational function is *to change people's mental and physical habits in regard to the forests.*"[42] For all the talk of organization, the forest-fire campaign eventually came down to the end of a matchstick.

What is most remarkable about the fire literature of the 1920s and '30s is that the issue was framed almost entirely by literary amateurs — by professional foresters and their close associates for whom the writing of stories and persuasive articles was mostly a sideline.[43] Fire control, for these authors and publicists, was not a matter of sentimentalism, but the core principle and key selling point of publicly funded forestry programs in the United States. In a sense, their crusade was disingenuous, because it sought to harness the romantic yearnings of the public to rally support for public policies that served the needs of industrial forestry. In the long run, the demonization of fire in the public mind would work *too* well; witness the outcry that accompanied foresters' decision to let parts of Yellowstone National Park burn in the 1980s. Fire, as it turned out, was too complex to be termed either friend or foe. It was, and will always be, a natural phenomenon whose presence is negotiated through structures of influence and power, ecological science and industrial need, and shifting ideals of what the forest is for.

CHAPTER NINE

Rural Zoning and the Synthetic Frontier

The cutover areas . . . speak as eloquently against haphazard development as any city condition. The spotting of these lands with remote or abandoned farms, . . . with resulting personal grief and social loss, [and] . . . the fire hazard . . . all cry out for planning, for social direction of individual effort.
— Wisconsin attorney general's opinion, 1931

On June 18, 1908, a young man named George S. Wehrwein stood before his classmates at the Oshkosh, Wisconsin, Normal School to deliver the valedictory address. The son of an immigrant farmer, Wehrwein possessed a keen and questing mind, and on this occasion he focused it on a question that had been troubling him for some time: the conservation of natural resources. His timing was auspicious. A month earlier, Theodore Roosevelt had summoned the nation's governors to Washington for a much publicized conference designed to place conservation at the top of the national agenda. With his customary thunder, TR had declared an end to "unrestricted individualism" and had offered in its place an appealing vision that linked conservation with the imperatives of efficiency and patriotism.[1] Wehrwein, just barely out of his teens, had honed his own speech through several drafts, and one can imagine him striving to replicate TR's bluster for the graduation crowd at Oshkosh. Rampant individualism, he declared, had resulted in man's failure to "read nature" and to discover its immutable laws. With foresight and planning by experts, the secrets of nature's "book" could be divined, Wehrwein believed, and famine and national ruin could be averted.[2]

A quarter century later, in 1933, a visitor called one day at the office of Louis Sorden, the agricultural agent for Oneida County in northern Wisconsin. The visitor, a would-be farmer, wanted to buy a parcel of land that had reverted to county ownership for tax delinquency. In days past, the county would have been delighted to sell the acreage to any person with good prospects; after all, the sale would put the land back on the tax rolls, where it would generate revenue for county government. But on this day, Sorden did something previously unthinkable: he said no.[3]

Sorden's decision was an outgrowth of rural zoning, a social experiment that capped a decade of legal-economic policy development for the cutover region. Authorized by the State of Wisconsin and implemented by individual counties, it allowed large expanses of cutover land to be designated as to best use: for timber growing, recreation, or general use as farms and home sites. Its most striking effect was to place vast areas of the North Woods off-limits to permanent human habitation — a restriction that could be enforced by the police power, if necessary. The aims were numerous: to prevent the wasting of labor and capital on poor agricultural land, to "block up" large tracts of land for forest development, to reduce the hazard of forest fires caused by isolated pioneers, and to save money on roads and schools in far-flung areas. As of mid-1937, northern and central Wisconsin counties had embraced zoning with great zeal, acting to bar settlement on more than five million acres.[4]

A chief architect of the rural zoning mechanism was George S. Wehrwein, who had risen from humble beginnings to become a professor of agricultural economics at the University of Wisconsin. In fieldwork and in the more rarefied realms of policy and theory, Wehrwein had spent his entire career making good on his youthful pledge to bring expert intelligence to bear on the workings of individuals. In the 1930s, Wehrwein also would become zoning's most prominent apologist, penning articles for popular and scholarly publications and speaking on radio to explain zoning's rationale and its benefits. In the process, he would reveal many of the ironies that were central to the cutover revolution. Rural zoning epitomized the double-edged qualities of much cutover planning: of good intentions coupled with paternalism, of the ascendance of articulate voices above the inarticulate, and of the elevation of societal goals above the freedom of the individual.

As one enthusiastic magazine article termed it, rural zoning would "revive the wilderness." But this new "wilderness" was essentially synthetic, having been invented by policy experts in the interest of furthering collective goals for the former frontier.[5] The remade forest would be a new and improved frontier, offering at least the illusion of wild land without the economic and social pathologies that were thought to accompany unregulated development. Natural in appearance, it was in fact the naturalistic product of a new planning ethos that, by 1933, had come to regard uncoordinated individual effort as wasteful and potentially ruinous. Its trees, managed by private and government foresters, would supply a steady and predictable stream of wood products for home and industry. But the landscape of the regrown "wilderness" was shaped as much by leisure as it was by the imperatives of industrial forestry. In tune with the forest-recreation movement, planners eventually would recognize the value of facilitating a sort of playacting in the forest — with camping, hiking, and other pursuits serving to refresh vacationing urbanites and to provide a touchstone to bygone frontier experience. But the frontier itself — defined as the unregulated encounter between man and nature — was gone for good, and no one among the New Deal planning generation appears to have mourned its disappearance.

Up from the Soil

George Wehrwein, more so than most of his colleagues, was well positioned to recognize the impact of social policy on individual Americans. He had moved up the academic hierarchy the hard way, working as a public-school teacher and as an extension worker in Texas and Washington state before gaining a Ph.D. degree at Wisconsin. When he joined the UW faculty in the late 1920s, his vitae boasted not only brief professorships at Northwestern and the University of Pennsylvania, but also several years of fieldwork during which he had witnessed farmers' struggles firsthand. Unlike some academics, such as his mentor Richard T. Ely, Wehrwein could empathize with the individual plowman even while shaping policies that were collective and often restrictive in nature. In the matter of farm tenancy, for example, Wehrwein fretted that landless men might become "careless and shiftless" without constant oversight. Absent any long-term tie to the soil, tenants might engage in destructive cropping practices, thus undermining the economic and social stability of the countryside. But Wehrwein's paternalistic outlook was

tempered by a genuine desire to see worthy farmers get ahead. The se-
cret, he believed, was to identify the prospective successful farmers from
among the masses of "shiftless, shifting, and economically irresponsi-
ble" tenants who were dragging down the quality of rural life. Policy
makers, Wehrwein asserted, had a duty both to identify the few talented
farmers and to devise methods to help them climb the "agricultural
ladder."[6]

Tenancy was not widespread in the Great Lakes states, but Wehrwein's
fertile mind soon drew parallels between the tenancy problem and the
emerging crisis in the cutover. While still a graduate student, he had
been apprenticed to Ely, who in the early 1920s was defining the terms
of land economics. The problem in the cutover was not so much one of
landless farmers as it was of farmers owning unsuitable land. Cutover
land was cheap, and its sellers were eager to unload it, so the farmer
who wanted land could probably get it. But many cutover soils would
not support the "carrying costs" of yearly taxes and interest, in addition
to providing a reasonable standard of living for a farmer and his family.
The tenancy problem — impressed upon Wehrwein during his work in
Texas — was one of purportedly substandard workers, while the cutover
crisis was one of substandard acres. Under Ely's tutelage, Wehrwein
would come to believe that both people and soils could be rank-or-
dered as to their fitness for agriculture. Wehrwein also would inherit
Ely's faith that the factors of agricultural production — labor, land, and
capital — could be manipulated via public policy to yield a productive
and socially stable landscape. As early as 1923, he was echoing his mentor's belief that the United States had too much land in crops and that
only a "retraction" in agricultural settlement would ensure prosperity
for the remaining farmers.[7]

Meanwhile, the cutover problem was forcing him to expand his outlook, beyond agricultural economics to the broader question of coordinating land uses for maximum economic and social utility. If the cutover would not produce food crops, it probably could produce trees
instead. Yet the question of cutover land utilization was far bigger than
potatoes or corn versus saw timber or pulpwood. P. S. Lovejoy, who
helped to launch the Michigan Land Economic Survey, conceived of
forest "products" in ever wider terms as the 1920s wore on, though he
was loath to acknowledge the purely spiritual or mythical value of the
woods. Expert planners were quite capable, for instance, of creating "big,

safe, green forest areas full of game, and streams full of fish."[8] Even in cases where forest uses could not be quantified, one of Lovejoy's colleagues asserted, planners still might achieve "the profitable use of every acre of land."[9] "Profit," in this case, included orderly development, prevention of waste, assurance of long-term productivity, and absence of social unrest. In tandem with this grand strategizing, government and university-based planners by 1930 had developed several workable legal-economic measures designed mostly to encourage timber growth on cutover acres that were submarginal for food crops. These included fire protection, abatement of property taxes on private forest lands, state financial assistance to local governments for the growing of public forests, and acquisition of tax-reverted lands by state and federal authorities for forest development.

Still, these measures could not address the fundamental economic crisis that was signaled by tax reversion. When lands were abandoned, local governments naturally scrambled to resell them. So long as the paradigm of "development" was construed in terms of agricultural settlement, submarginal lands seemed destined to go through a continuous cycle of sale, farming, failure, and abandonment, with the cost of local services to settlers often exceeding the meager tax revenues generated by these poor acres. Rural zoning — with the promise of removing large tracts from permanent habitation — held the potential to rationalize land use by confining crop agriculture to areas where it was most likely to pay. Submarginal acres then could be devoted to forestry under private and public ownership, with state and federal governments underwriting the long-term costs of growing tree crops or furnishing recreation havens for visiting urbanites. In addition, zoning offered a ready solution for what planners viewed as a particular social pathology among isolated cutover farmers. In the words of one forester, a properly administered woodland could produce not only trees and recreation, but a "sustained annual yield of prosperous, healthy, and happy human beings."[10]

The "Pioneer Fringe" and the "Receding Frontier"

For planners, the economic crisis that prompted cutover zoning was a familiar and daunting litany. The difficulties of cutover agriculture — exacerbated by the farm depression of the 1920s — had left the region dotted with abandoned farms and idle land. The farms that remained often were isolated and located on poor soils. Because huge areas of land

were idling in tax delinquency, local governments were starved for revenue. Against this dismal backdrop, costs of providing government services to cutover settlers were unusually high, because there were fewer
people per mile of road or per public-school teacher than in more populated areas. (School transportation was a problem, too. One study of
twenty-eight isolated farms in Minnesota found that the cost of getting
the settlers' children to school far exceeded the families' entire property
tax payments.[11] In northern Wisconsin, it was not unusual for cutover
schoolchildren to board with families in town during the winter, with
local governments paying the bill.) Without exaggeration, Lovejoy referred repeatedly to cutover governments being "practically bankrupt."
As of 1928, for example, fifteen Michigan counties each received more
than twice as much money from state aids as they paid to Lansing in
taxes.[12]

Recognizing the crisis was far different from resolving it. With large
portions of the cutover apparently destined for public ownership, it remained for Wehrwein and his colleagues to devise ways of reviving it
and to sell those ideas to the academic community, to political and business interests, and to the general public. Wehrwein recognized that much
of the groundwork would necessarily take the form of tutorials, in which
he would outline the fundamentals of land economics for a nonspe-
cialist audience.

Perhaps the toughest task was selling the idea of centralized planning
to a citizenry that had been steeped in the myth of rugged individualism. In an early radio talk on University of Wisconsin station WHA —
which was to become a regular forum for him throughout the 1930s —
Wehrwein grappled with the question at the heart of all land-use issues:
What was land for? Perhaps harking back to his Normal School speech
more than twenty years before, he declared that land existed to serve
man, not the other way around. To employ lands in ways that wasted
human effort was not only inefficient, it was contrary to discernible laws
of economics: "Every acre of our country has a real use, if we will only
take the pains to discover it and direct it into that utilization." Northern
agriculture might provide a subsistence living, but lands that yielded
only a meager return did not belong in farms at all: "[T]he land keeps
the farmer going even after the money profits are gone. However, if land
is brought into use, or kept in use at the expense of a decent return on
the investment or at the sacrifice of an adequate standard of living avail-

able in other callings, the land is no longer serving man but man is being sacrificed merely to keep land in a use which is not the right use."[13]

That simple pronouncement sprang from years of intensive study. Most notably, Wehrwein had been heavily influenced by Johann Heinrich von Thünen, whose *Der Isolierte Staat* (The isolated state) he had read in its original German. Von Thünen's book, published in 1826, codified land use in geographical terms and suggested that a "correct" use existed for each point within the hypothetical state. Von Thünen suggested that empirical study would support the theory by revealing that specified inputs of capital, land, and labor within selected areas would yield predictable outputs. His theory, in short, was the germ of land economics, a century before the discipline was formally developed. Wehrwein knew better than to invoke the name of a long-dead German intellectual during his radio chats with Wisconsin farmers. But von Thünen's work had convinced him that, with enough data and study, the "right" use for every acre could be discerned.[14]

Another key influence was the geographer Isaiah Bowman, whose book *The Pioneer Fringe* appeared in 1931. Bowman, like Wehrwein, saw a gross imbalance between human aspirations and the realities of land use. During the 1920s, the Malthusian specter of famine had been washed away by successive waves of agricultural surpluses. Yet Americans — infused with nineteenth-century ideals of pioneering yeomanry — continued to push the frontiers of settlement in the cutover and in the West. A rising standard of living in the United States meant that the new pioneers were likely to harbor unrealistic expectations of quick prosperity, Bowman believed. And they were certain to demand that governments provide amenities — particularly roads and schools — that their forebears had built on their own. Twentieth-century pioneers were not rustics or hermits; they were full citizens who made sacrifices in the present in hopes of future gains. The problem, as Bowman saw it, was that many of them were swimming against the tide of economic reality. The workings of the market would dash their illusions soon enough, but Bowman saw this process as enormously wasteful. Aggregate data and university expertise could be brought to bear on the problem through the workings of the state, with the aim of discouraging agriculture in areas where it would not pay. In other words, no farmer should sweat a drop in pursuit of a futile endeavor. Bowman called for a new "science of settlement," one that would encourage pioneering on selected lands but discourage or even forbid it in many others. "It is unintelligent to grow everything that can be grown in a given place," he declared.[15]

Even as Bowman wrote, Wehrwein's thinking was following a parallel track. Wehrwein recognized the longtime American emphasis on "town building." Pioneers, he knew, traditionally had sacrificed material comforts and had put themselves at the mercy of nature, with expectations that communities eventually would spring up around them. Four decades after the closing of the frontier, the cutover seemed to hold out new hope for individualism and an outlet for those whose ambitions had been frustrated in the cities. The problem, as Wehrwein saw it, was that the cutover pioneers had hit a wall of economic limits. With the tax-reversion crisis, farms were being abandoned, not built; land was yielding to brush, not to the plow. The frontier, in other words, was not advancing. It was receding. This single fact turned the idea of town building into a cruel hoax and threatened to unleash a host of economic and social problems: "The psychology and social structure of rural life on a receding frontier are different from those on an advancing frontier. Institutions are dying instead of being born. It has been observed in Canada that when the retreat begins an inverse natural selection leaves behind the most shiftless and inefficient, isolation increases, and a notorious rural slum is created."[16]

Government, Wehrwein asserted, had a duty to "maintain a satisfactory rural life in the marginal areas." It could do this by guiding settlers to lands most likely to support agriculture and by forbidding settlement in areas of poor soil. The locus of decision making, formerly located with the individual, would shift to the state. Some lands were destined for government ownership. These submarginal acres — fit only for forests — already were making themselves known through the workings of the marketplace. Some of them had reverted for nonpayment of taxes several times. Governments might try desperately to resell these lands to keep them on the tax rolls. But by refusing to take title to the lands, public officials were merely suspending them "in mid-air between private and public ownership." Only reforestation, combined with such measures as fire protection, would allow these acres to yield the economic rents they were capable of, Wehrwein believed.[17]

Wehrwein, more so than Ely, tended to think that the "right" use for land was clearly discernible. At times he spoke of land-use questions in religious terms, saying that gross imbalances such as the cutover crisis

George S. Wehrwein and a colleague, Mary L. Shine, at work at the University of Wisconsin, about 1927. Photograph courtesy of State Historical Society of Wisconsin, WHi (X3) 46598.

resulted from human beings' failure to heed the laws of nature, thus of God. Beginning about 1930, however, he occasionally acknowledged that ideas of land use — such as the boundary between "good" agricultural lands and "poor" ones — were social and political constructions, subject to change over time. Such an admission, though, did not stop him from issuing blunt advice to people who asked his opinion about farming in the cutover. Wehrwein was alarmed by a burst of "back to the land" enthusiasm in the early years of the Great Depression. Jobless city folks with visions of self-sufficiency could only worsen the crisis in the North, he believed. When a Milwaukee-area man inquired about farming 190 acres of cutover land he owned in Iron County, Wisconsin, Wehrwein's advice was unequivocal: Don't. Wehrwein noted that many unemployed people had moved into abandoned farms in the North, but he added that their fate probably would be grim: "I am afraid that they have substituted slow starvation for quick starvation, unless they can find work for the winter." He advised his correspondent to stay home and plant a vegetable patch instead.[18]

An Idea Becomes Law

Why did Oneida County, with the assistance and encouragement of the State of Wisconsin, begin turning away prospective farmers in 1933? The reason was simple arithmetic. Wisconsin law required that all settlers be provided with roads and schools. In many cases, tax-delinquent parcels were so remote that the county would spend far more than it ever could hope to receive in tax revenue. County officials doubtless remembered the occasion when $1,200 had been spent to build a road to an isolated pioneer. The road had been used just once—when the farmer moved out.[19]

Wisconsin was the first state to provide for rural zoning, and its experience became a model for other states. A series of enabling laws gave counties broad powers to zone for agriculture, recreation, and forestry uses. In the cutover, zoning could be used to "block up" large tracts of forest lands. Existing settlers could remain on these lands, but no new settlement would be permitted. By restricting settlement to small, contiguous areas, counties could lower their costs of providing government amenities, particularly roads and schools. Restrictions on new settlement also would reduce the incidence of human-caused forest fires and prevent isolated folk from violating the game laws, a temptation that—at least according to the planners in Madison—was omnipresent in the woods. Encouraged by UW land economists and extension workers, Oneida County enacted the first county zoning ordinance in 1933. By the early 1940s, twenty-six more counties in northern and central Wisconsin had followed suit.[20]

University of Wisconsin officials took direct responsibility for "selling" the zoning plan county by county. "An enlightened public opinion, conscious of the purpose and operation of a zoning ordinance, will result in continued effectiveness of this instrument of constructive control," they declared in 1934.[21] In some counties, public meetings on the zoning question revealed much local scorn for cutover settlers, who were thought to be degrading the community with their antisocial ways and dependence on government relief. A Price County lawyer, for instance, suggested ominously that zoning would be an ideal mechanism for removing troublemakers from the "back 40s."[22] In Oneida County, the record is remarkable only for the near absence of opposition voices. The *Rhinelander Daily News* supported zoning foursquare, noting that it would forestall "the breakdown of social relationships" involved in isolated settlement. At a public hearing in May 1933, one citizen objected to the plan, saying that it ignored the "sentimental values attaching to property."[23] Another resident's complaint elicited a lecture about the new realities of interdependence in woodland communities:

"Do you mean to tell me that you are going to tell a free-born American citizen that he can't do with his land as he pleases?" was a direct question at one of the meetings. Yet, after this man was shown that, when a "free-born American citizen" in doing what he pleases burdens the taxpayers with an unnecessary school, a road which the settler might use merely to move out again, to say nothing of the fact that the town might have to feed him, he was convinced of the soundness of zoning principles.[24]

After a series of town meetings conducted by UW experts, the Oneida County Board of Supervisors approved the plan on May 16.[25]

The "Tragedy" of Individualism

To George Wehrwein, zoning made perfect sense. In his view, pioneers were constantly pushing at the margins of settlement, going where few had gone before in hopes of long-term gains. This was the spirit of "town building," and it had worked admirably in places such as the Ohio Valley. But the harsh nature of cutover agriculture meant that few of these twentieth-century pioneers would succeed. In most cases their toil simply was wasted. Instead of having towns spring up around them, these people would find themselves increasingly isolated, vulnerable to all means of economic and social ills on the "receding frontier." So governments would step in and determine the margins of feasible settlement themselves. After 1933, the cutover frontier would be collectively defined, and no one would be allowed to push its boundaries. In Wehrwein's eyes, eager settlers were not being thwarted, but rather were being protected against the consequences of their unknowing enthusiasm. Zoning, he said in a 1935 radio talk, "will prevent the recurrence of the tragedies of the past."[26]

Implicit in this argument was the idea that state expertise could help people make better choices than they would make on their own. Wehrwein's motivations in this regard were public-spirited. In his commencement speech at the Normal School a quarter century before, he had spoken confidently of scientists' ability to decode the book of nature and to share its secrets with the broader populace. (The people's greatest need, he had stated, could be summed up by the enigmatic last words of Goethe, "More light!") Wehrwein erred, though, in assuming that all people should share his values as to land use. In particular, he appears to have believed that pioneering was outmoded and that the only true measure of a farm's success was its income from cash crops. The "science of settlement" (as Isaiah Bowman put it) clearly told people that it was sensible to live in some places and not in others; why wouldn't the people listen? Wehrwein sympathized with a colleague in Idaho who complained about people living in the mountains when a better standard of living could be had on nearby farms. "I suppose we will just have to let people live where they please," Wehrwein told this colleague resignedly, "but from the standpoint of furnishing schools, roads and other things through public expenditures, the story is another one."[27]

At his worst, Wehrwein blamed cutover settlers for their own problems. Many of them, he said, harbored fanciful illusions of becoming "second Daniel Boones. The trouble is they are not willing to accept the kind of life Daniel Boone was willing to live." The cutover pioneers could be fickle. Having fled civilization, they yearned for its amenities: "They want schools for their children, and a road to get to market, and they want the road plowed out in winter besides." Fascination with wilderness life was a temporary fixation, Wehrwein believed. Once it had worn off, the backwoods people would be "perfectly willing to let the town or county support them." Wehrwein's theories about social decay on the "receding frontier" were never definitively tested during this period. But it seems clear that he was willing to accept many of the assumptions about the rise of "rural slums" in the North.[28]

Because rural zoning was not retroactive, some settlers remained in areas that were designated for forestry or recreation. Beginning about 1933, Wehrwein and his students spent thousands of hours enumerating these "nonconforming" land users, as they came to be known. Numbering only about two thousand, these Wisconsin pioneers were widely dispersed in the woods. Data from a 1933 survey showed that, on average, they lived almost a mile from their nearest neighbor. In Langlade County, the average isolated settler had to journey fifteen miles to reach his church. Annual cash income from pioneer farms averaged just $147. Some pioneers clearly had harbored expectations of town building that had failed to bear fruit. One family, having toughed it out for thirteen years, lamented the fact that they still had to board their children in town with strangers during the school year: "We . . . thought that by this time there would be enough settlers to build a school here." Complained another: "We have no community."[29]

The validity of such complaints notwithstanding, Wehrwein and his crew approached their work with preconceptions that were bound to distort their findings. Chief among these prejudices was the idea that living in the woods was fundamentally illogical. An enumerator working in Douglas County in 1934 wondered whether he should fill out schedules for squatter families who were "more or less undesirable and shiftless." Another, in Marathon County, wrote off the potential of nonconforming settlers there with a single sentence: "Many of them are purely town charges and wouldn't work if they had a good farm."[30] Wehrwein himself fretted that, without neighbors to watch over them, people in the forest might revert to a sort of savagery, bereft of any standards of morality or cleanliness. He quoted one observer who had described the cutover as a nightmarish frontier hell: "Sanitary conditions are something terrible; most of the settlers are poorly educated, poorly nourished, poorly housed, and as the result of generations of eugenic carelessness, they lack the fiber to do anything for themselves. Many of the women are on the point of insanity; if something should snap they would be hopelessly deranged. The way they live is bound to breed degenerates."[31]

Toward the end of 1939, a team of authors lauded the Wisconsin zoning experiment. In an article reprinted in the *Reader's Digest*, they declared that the existence of submarginal farms had led to countless social "evils," including wasted effort and welfare dependence. Just a few years before, many northern counties had been "on the verge of bankruptcy." But zoning had allowed communities to "control land use for the greatest common good." The result had been a quick and irrefutable economic revival, even in the midst of the Great Depression. Three-quarters of the land zoned for forestry already was growing trees again. The northern counties had become "financially sound," and tax delinquency had been reduced to "a minor irritation." Never again would the region host dreams of agricultural splendor, of stumps yielding to the plow. Instead, cutover planning had laid the foundation for a new world in the North: one of forestry and tourism, in which every acre would yield some sort of benefit, whether in the form of timber, tourist dollars, or simply a breath of pine-scented air. By 1940, the cutover was back on a paying basis.[32]

But at what cost? It is clear that George Wehrwein's vision for the cutover had no place for subsistence farming. By calculating the inputs and outputs of the economy in material terms, the land economists effectively squelched debate on the larger meaning of yeomanry in the North.

Farmers who did not produce crops for market were deemed "failures," and their efforts were judged a "waste," simply because these people could have enjoyed a higher material standard of living by doing something else. Through rural zoning, the land economists had furthered a revolution that had been launched a quarter century earlier by their colleagues in agronomy. By the mid-1930s, subsistence farming would be viewed as an archaic form of agriculture that had no place in an integrated, externally directed rural economy. Unlike the agronomists, who had encouraged production of cash crops at the margins, the land economists focused instead on "eliminating farmers who threatened American economic viability and political stability."[33] The pioneer was dead, and the hand that guided the plow now belonged to the paternalistic state.

As Wehrwein himself came to admit, the line between "good" and "bad" agricultural lands was not drawn by the hand of God (as Wehrwein had fancied in 1908), but by fallible human beings based on shifting human values. The drawing of this line, as exemplified by rural zoning, was not so much a democratic exercise as it was a triumph of the university-based "experts." Many cutover farmers were immigrants; many others were unschooled in the ways of political persuasion and media image-making. Doubtless many of them believed that their opinions mattered little in the face of university expertise and the bewildering machinery of the state. (The total absence of American Indian voices in the cutover literature adds weight to this contention.) Historians must be careful not to romanticize the independent farmer or to underestimate the cutover planners' contribution to the overall good of the once-and-future North Woods. The cutover revival was not a zero-sum game; in fact it can be argued that far more was gained than was lost. But the gains have been amply documented; the losses have not. Historians of the cutover are just beginning that task.

Epilogue

In 1936, economist Rexford Tugwell sounded the death knell for the myth of American pioneer independence. Tugwell, a member of Franklin D. Roosevelt's "brain trust," declared flatly that individualism in land use was dangerous to the national welfare. He asserted that nature itself abhorred the absence of planning. He cited Wisconsin's rural zoning laws as an example of how collective imperatives could, and should, restrain the former pioneer. Tamed by community pressure, the individual "will have to learn to cooperate . . . he will have to conform."[1]

Tugwell's pronouncement spoke to a revolution in the methods and ideas of land use, a revolution that was felt especially strongly in northern Michigan, Wisconsin, and Minnesota. By the early 1930s, the Great Lakes cutover had been tamed and groomed. Formerly a landscape given to wild excesses, the cutover henceforth would come under the technocratic oversight of downstate experts, who would manage it from afar to produce crops as varied as timber, fish and game, and scenery. After the reckless abandon of the logging era and the troubled experiment with agriculture, the economy of the North would settle into a stable if unspectacular pattern, with tourism, government, and wood-products industries supplying most of the region's jobs. Though maintaining at least an appearance of its wild past, the naturalistic northern landscape would epitomize the sort of continuously productive nature that had been the stuff of planners' fantasies since the Progressive Era. A few years after Tugwell's decree, one informational bulletin could state confidently that planners had a self-evident "duty" to regulate the use of land "according to its best adaptation." For decades to come, the ability of experts to discern and dictate that "best" use would scarcely be questioned.[2]

The peculiar historical circumstances of the Great Depression would consummate the cutover revolution sooner than anyone might have dreamed. Economic troubles facilitated the continued transference of northern land into government hands through tax reversion and also the political consensus for large public employment projects — notably the Civilian Conservation Corps — that would help create the physical structure of the "reimagined" public forest. The depression's ecological calamity — farmland erosion, drought, and the resulting Dust Bowl — would elevate questions of resource use to the status of national emergency. Tugwell was not the only American to regard the Dust Bowl as a swirling omen of nature, a wrathful warning that the folly of individualism had come to "an inglorious end."[3]

The revolution was far-reaching, yet fundamentally conservative. Under the aegis of technocratic Progressivism, it would imbue every resource (including human labor) with public purpose, channeling those resources into uses that were thought to contribute to national wealth, economic predictability, and social stability. Historian David B. Danbom writes that the revolution in agriculture and in wider land-use questions represented "an urban attack on rural isolation, individualism, and self-sufficiency," which sought to make rural America "a socially and economically organized and efficient part of an increasingly interdependent nation."[4] Writers and editors, to varying degrees, tended to adhere to this conservative vision. Even their prescription for preserving remote tracts of "wilderness" as an arena for strenuous pioneer experience was more in keeping with an emerging therapeutic culture than with any comprehensive scheme for remaking rural life.

As in agriculture, the cutover revival had its nostalgics, who hoped that the gospel of efficiency and planning might somehow regenerate an indigenous, independent, stable way of life among the region's permanent inhabitants. But a lopsided power relationship dictated otherwise. Because the money, political clout, and knowledge involved in remaking the forest lay in the cities and the universities, the new woodland landscape would come to reflect the desires and dreams of the more cosmopolitan population. The untamed forest would give way to the industrial timber reserve, the taking of fish and game for subsistence would be supplanted by regulated harvest under the sporting ethic, and the pioneer cabin would yield to the tourist lodge. The cutover nostalgics would discover that it was impossible, in Danbom's words, "to save the past with the tools of the future."[5] The depression-era cutover constituted a vast laboratory where social scientists and technical professionals might have imagined an entirely new rural culture. Instead it became, arguably, something less — a landscape dedicated to serving the needs and easing the social stresses of the existing industrial order.

Those needs and stresses, and their solutions, would prove more mutable than anyone knew, however. Samuel P. Hays has described how the rising affluence of post–World War II America — signaled by a shift in social emphasis from efficiencies of production to "quality of life" — has come to be reflected in a new environmental awareness. In the consumption-oriented "age of amenities," Hays writes, clean air, clean water, and open spaces have become yardsticks of an acceptable standard of living. In his view, the environmental impulse has not been an obstacle to progress, but "an expression of deeply rooted human aspirations for a better life."[6] Rather than scorning technology, it has goaded scientists into attacking the most intractable environmental problems of recent decades: radioactivity, chemical pesticides, toxic pollution, and transnational phenomena such as acid rain and global warming. And far from eschewing "the system," it has spun its own networks of expertise and influence to bring about the environmental amenities that can be secured only through the large-scale machinery of the state, such as innovative regional planning and protected wildlife areas.

The present-day North Woods is no pristine paradise, much as tourism promoters would like people to think it so. It is a heavily used environment, beset by problems ranging from noise pollution to pernicious lakeshore development. The struggle to keep it livable depends, as environmental movements often do, on employing state authority to restrict private actions that have disproportionate social costs. State sponsorship is no guarantee of an idea's environmental soundness, either, especially in cases where immediate desires clash with more ethereal goals such as biodiversity, which have not yet gained wide cultural currency or political salience. Much as it was three-quarters of a century ago, the quest to define the woods is an ever shifting calculus of social priorities and networks of influence, a game whose rules are constantly changing and whose outcome is never certain. The new "wilderness," if one chooses to call it that, is not so much "God's country" as it is a culturally crafted place that is utterly dependent on the energies of human beings.

Trees, in other words, will continue to grow much as they always have, but the metaphors they evoke almost certainly will keep changing. Over time, the terms of environmentalism have been extraordinarily malleable, which may in fact constitute the movement's central beauty. Aldo Leopold once wrote that the chief duty of the "recreational engineer" was not to devise physical structures but "to promote perception" — that is, to school the public in the art of appreciating nature. Singly and in community, Leopold wrote, people possessed a "mental eye" that governed what they might see when looking at a forest, a lake, or a mountain. The policy maker (or writer) who could shape that vision might discover that supposedly "finite" resources were almost infinitely expandable, as with a woodland that could accommodate many people if they would only respect it and protect it and tread lightly under its sheltering branches. "Recreational development is a job not of building roads into lovely country, but of building receptivity into the still unlovely human mind."[7]

Even when dedicated to profoundly conservative purposes, the American forest has always displayed an ability to slip those bonds and to scatter seeds of myth, mysticism, and — within limits — the sort of individualism that allows the wandering spirit to tolerate the aggravations of living in community. Its continued making will rest not just with the technical forester, but with the activist, the ethicist, the ordinary citizen — and, of course, with the writer bold enough to "reimagine" and reconsecrate the face of the Earth.

Notes

Introduction

1. Theodore Roosevelt Jr., "Fishing in Wisconsin," *Scribner's Magazine* 77, no. 5 (May 1925): 463–69.

2. Ibid., 463, 466.

3. See Guy Alchon, *The Invisible Hand of Planning*. Alchon describes how macroeconomic planners in the 1920s sought a regulated middle way between monopoly capitalism and socialism, with the aim of making capitalism "continuously productive," countering the swings in the business cycle and buffering the cruelest excesses of unbridled free enterprise. The hero of the movement, Alchon writes, was the "technocratic professional."

4. J. Robert Barth, S.J., "Theological Implications of Coleridge's Theory of Imagination," in Christine Gallant, ed., *Coleridge's Theory of Imagination Today*, 3–13. Samuel Taylor Coleridge defined two levels of imagination: the "primary," which allowed ordinary people to find meaning and order in an otherwise chaotic world, and the "secondary," which was a higher imagination of the sort used by artists. The artist "breaks down, 'dissolves,' the unity he has perceived in and among the natural shapes of [nature] . . . to allow a new chaos to emerge, as it were, and then to shape a new unity out of his own consciousness of it, expressing it in water colors or plaster or the sounds of music." Barth, "Theological Implications," 4. For the concept of "reimagining" I am indebted to Father Michael Himes of Boston College.

5. P. S. Lovejoy, "The Promised Land: In the Northern Cut-Over Country Lies America's Great Ungrasped Opportunity," *Country Gentleman* 85, no. 51 (December 18, 1920): 3–4, 32.

6. P. S. Lovejoy, "Suckers, Grazing, Farming or What? For Our 228,000,000 Acres of Desolate, Fire-Swept Cut-Over Land," *Country Gentleman* 84, no. 28 (July 12, 1919): 3–4, 32, 34. The acreage figure of the title was Lovejoy's estimate for the entire United States.

7. See, for example, Susan L. Flader, ed., *The Great Lakes Forest*. Earlier studies tend to focus narrowly on the scientific, legal, or policy questions involved in re-

making the forest. These works include Vernon Carstensen, *Farms or Forests;* Arlan Helgeson, *Farms in the Cutover;* and Erling D. Solberg, *New Laws for New Forests.*

8. For the quality-of-life ideal, see Samuel P. Hays, *Beauty, Health, and Permanence.*

9. William Cronon, "The Trouble with Wilderness; or, Getting Back to the Wrong Nature," *Environmental History* 1, no. 1 (January 1996): 7–28.

10. See, for example, Gray Brechin, "Conserving the Race: Natural Aristocracies, Eugenics, and the U.S. Conservation Movement," *Antipode* 28, no. 3 (July 1996): 229–45. The critical approach to cutover history is epitomized by Robert Gough, *Farming the Cutover.*

11. Michael Pollan, *Second Nature,* 171.

1. "Timber Famine," the Quest for Production, and the Lingering Frontier

1. Michael Williams, *Americans and Their Forests: A Historical Geography,* 372–73.

2. Thomas R. Cox, Robert S. Maxwell, Phillip Drennon Thomas, and Joseph J. Malone, *This Well-Wooded Land,* 158. A "board foot," the basic measure of standing timber or raw lumber, is a piece of wood one foot square and one inch thick. Timber cruisers and foresters sometimes referred simply to "feet," meaning board feet.

3. Hazel H. Reinhardt, "Social Adjustments to a Changing Environment," in *The Great Lakes Forest,* ed. Susan L. Flader, 205–19.

4. Charles E. Twining, "Plunder and Progress: The Lumbering Industry in Perspective," *Wisconsin Magazine of History* 47, no. 2 (winter 1963–64): 116–24.

5. George P. Ahern, "Deforested America: Statement of the Present Forest Situation in the United States," U.S. Senate Document no. 216 (1929), in box 553 of the Gifford Pinchot papers, Library of Congress, Washington, D.C. (hereinafter referred to as Pinchot papers).

6. Henry S. Graves, "A National Lumber and Forest Policy," USDA Circular 134, April 1919, in box 562, Pinchot papers.

7. John Milton Cooper Jr., "Gifford Pinchot Creates a Forest Service," in *Leadership and Innovation,* ed. Jameson W. Doig and Erwin C. Hargrove, 63–95.

8. "Forest Devastation: A National Danger and a Plan to Meet It," *Journal of Forestry* 17, no. 8 (December 1919): 911–45. Pinchot's vision was several degrees more radical than that of Graves, who sought mainly state control of forestry rather than federal and retained faith in a model of cooperation, not coercion. Williams, *Americans and Their Forests,* 444–45.

9. Gifford Pinchot, "Time to Compel Saving Trees," *San Francisco Examiner,* February 11, 1920, clipping in box 562, Pinchot papers.

10. Gifford Pinchot, "The Lines Are Drawn," *Journal of Forestry* 17, no. 8 (December 1919): 899–900.

11. Harold T. Pinkett, *Gifford Pinchot: Private and Public Forester,* 78.

12. Ralph W. Hidy, Frank Ernest Hill, and Allan Nevins, *Timber and Men,* 297.

13. Carl E. Krog, " 'Organizing the Production of Leisure': Herbert Hoover and the Conservation Movement in the 1920's," *Wisconsin Magazine of History* 67, no. 3 (spring 1984): 199–218.

14. Stephen Ponder, "Gifford Pinchot: Press Agent for Forestry," *Journal of Forest History* 31, no. 1 (January 1987): 26–35; and Stephen Ponder, "Federal News Management in the Progressive Era: Gifford Pinchot and the Conservation Crusade," *Journalism History* 13, no. 2 (summer 1986): 42–48.

15. "Forest Devastation," 935.

16. Russell Watson, "Forest Devastation in Michigan: A Study of Some of Its Deleterious Effects," *Journal of Forestry* 21, no. 5 (May 1923): 425–51.

17. Richard H. Stroud, ed., *National Leaders of American Conservation*, 77–78.

18. Ovid M. Butler, speech text of October 7, 1919, in Ovid M. Butler manuscript file, Forest History Society, Durham, N.C. The Forest Products Laboratory was established at Madison in 1910. It employed as many as 460 people during the world war and about 200 after the armistice. It investigated the physical properties of wood, methods for drying and preserving lumber, pulp and papermaking methods, development of new wood products such as plywood and pressboard, and derivation of a wide range of allied products, such as turpentine and naval stores, through wood chemistry.

19. The magazine would, at various times, be called *American Forestry, American Forests*, and *American Forests and Forest Life*. The American Forestry Association (AFA) was a public membership group with a long history of lobbying for forest conservation. It should not be confused with the Society of American Foresters, which was a professional organization. The AFA's 16,000 members in 1920 included foresters, landscape architects, interested members of the public, and timber executives. Alexandra Eyle, *Charles Lathrop Pack*, 133–66.

20. Ovid M. Butler, "Our Forest Hunger," *American Forestry* 29, no. 349 (January 1923): 3–13.

21. Gifford Pinchot, "The Blazed Trail of Forest Depletion," *American Forestry* 29, no. 354 (June 1923): 323–28, 374.

22. Ellis Hawley, "Three Facets of Hooverian Associationalism: Lumber, Aviation, and Movies, 1921–1930," in *Regulation in Perspective*, ed. Thomas K. McCraw, 95–123. Hawley says Hoover placed excessive faith in technical fixes, such as standardization of lumber grades, which were not sufficient to prevent "wastage and heightened social costs" as timber cutters struggled for survival in the 1920s. Ibid., 108.

23. Ovid M. Butler, "Henry Ford's Forest," *American Forestry* 28, no. 348 (December 1922): 725–31.

24. Williams, *Americans and Their Forests*, 429.

25. See, for example, "Built-Up Wood," *Literary Digest*, February 7, 1920, 30–31, in Ovid M. Butler publications file, Forest History Society, Durham, N.C. An exemplar of the new wood science could be found in an airplane propeller: glued up from pieces of wood, it had to be exceptionally strong, durable, and resistant to dampness and bacterial invasion.

26. Aldo Leopold, "Game Management in the National Forests," *American Forests and Forest Life* 36 (July 1930): 412–13.

27. O. M. Butler, "The Profession of Forestry and the Public Mind," *Journal of Forestry* 23, no. 5–6 (May and June 1925): 446–50.

28. "Planting Trees to Serve Our Children," *World's Work* 46, no. 1 (May 1923): 60–68.

29. Roland Marchand, *Advertising the American Dream*, 9.

30. Frank A. Waugh, "What Is a Forest?" *Journal of Forestry* 20, no. 3 (March 1922): 209–14. European forests had long been managed for wild game and recreation — but only for use by nobles, Waugh noted.

31. Henry Nash Smith, *Virgin Land*, 293.

32. Pinkett, *Gifford Pinchot*, 78.

33. Gifford Pinchot, *Breaking New Ground*, 259.

34. P. S. Lovejoy, "In the Name of Development," *American Forestry* 29, no. 355 (July 1923): 387–93, 447.

35. Frederick Jackson Turner, "The Significance of the Frontier in American History," in *The Frontier in American History*, 2–3, 22. It is important to note that Turner placed no great value in wilderness itself, but rather in the zone where white man and wilderness met. The *process* of taming the wild (and its Indian inhabitants) determined the American character, Turner believed.

36. Richard Slotkin, *Gunfighter Nation*, 22–23.

37. Owen Wister, *The Virginian*, 147.

38. Frank Norris, "The Frontier Gone at Last," in *The Responsibilities of the Novelist and Other Literary Essays* (New York: Doubleday, Page, 1903), 69–81.

39. David W. Levy, *Herbert Croly of "The New Republic,"* 101. For an explication of the tensions between self-interest and the greater good, see James T. Kloppenberg, *Uncertain Victory*.

40. Warren I. Susman, *Culture as History*, 35.

2. The Agricultural Problem and the Land-Economics Solution

The epigraphs are from H. L. Russell, "Farms Follow Stumps," University of Wisconsin Agricultural Experiment Station Bulletin 332 (April 1921), 3; and Richard T. Ely, "Farm Homes and Our National Welfare," speech delivered at the "Farm City Conference," New York, February 3, 1922, in box 4 of the Richard T. Ely papers (additions), State Historical Society of Wisconsin, Madison (hereinafter referred to as Ely papers).

1. Ellis W. Hawley, *The Great War and the Search for a Modern Order*, 26.

2. Robert J. Gough, "Richard T. Ely and the Development of the Wisconsin Cutover," *Wisconsin Magazine of History* 75, no. 1 (autumn 1991): 3–38.

3. Richard T. Ely and Edward W. Morehouse, *Elements of Land Economics*, 48 n. 1.

4. Jacob Perlman, "The Recent Recession of Farm Population and Farm Land," *Journal of Land and Public Utility Economics* 4, no. 1 (February 1928): 45–58.

5. Harry C. McDean, "Professionalism, Policy, and Farm Economists in the Early Bureau of Agricultural Economics," *Agricultural History* 57, no. 1 (January 1983): 64–89. The term "racial" in this context probably referred to human genetic stock.

6. Even before the 1920s, cutover agriculture faced tough sledding. In its 1922 *Yearbook*, for example, the USDA reported that the number of cutover farms in Michigan had actually fallen from 1911 to 1920, despite aggressive boosterism by state and local authorities. Norman J. Schmaltz, "The Land Nobody Wanted: The Dilemma of Michigan's Cutover Lands," *Michigan History* 67, no. 1 (January/February 1983): 32–40.

7. Edward H. Beardsley, *Harry L. Russell and Agricultural Science in Wisconsin*, 88.

8. Richard T. Ely, "Private and Public Colonization; or, Organized Settlement of the Land," speech for Wisconsin Association of Real Estate Boards, Milwaukee, February 20, 1923, in box 9, Ely papers.

9. *Minnesota: Its Resources and Progress; Its Beauty, Healthfulness and Fertility; and Its Attractions and Advantages as a Home for Immigrants, with a Descriptive Map* (Saint Paul, Minn.: Press Printing Company, 1870), 68, in subject file "Minnesota," Forest History Society, Durham, N.C.

10. Arlan Helgeson, *Farms in the Cutover,* 26.

11. Ibid., 29–31.

12. John I. Kolehmainen and George W. Hill, *Haven in the Woods,* 67–68.

13. Hazel H. Reinhardt, "Social Adjustments to a Changing Environment," in *The Great Lakes Forest,* ed. Susan L. Flader, 205–19.

14. Lucile Kane, "Settling the Wisconsin Cutovers," *Wisconsin Magazine of History* 40 (winter 1956–1957): 91–98.

15. E. L. Luther to K. L. Hatch, February 7, 1912, in box 1 of the E. L. Luther papers, State Historical Society of Wisconsin, Madison (hereinafter referred to as Luther papers). Hatch was head of the UW's Agricultural Extension Service and father of the Hatch Act, which provided federal money to state agricultural experiment stations.

16. E. L. Luther to K. L. Hatch, February 28, 1912, box 1, Luther papers.

17. *Rhinelander (Wis.) News,* December 9, 1910.

18. Faast's contact with UW officials was a little *too* close. Agriculture Dean Harry Russell helped found Faast's company in 1917 and eventually would invest $35,000 in it. Russell recognized this potential conflict of interest and tried to separate his personal investments from his professional work. But his stake in colonization at least partially blinded him to the problems of cutover farming. Beardsley, *Harry L. Russell,* 130–31.

19. Ben F. Faast, "The Real Way for Colonization," *National Real Estate Journal,* n.d. [1917], in box 63, Ely papers.

20. Richard T. Ely, discussion notes for "Eau Claire Conference on Land Settlement," September 24, 1918, box 63, Ely papers.

21. Schmaltz, "The Land Nobody Wanted," 34, 37.

22. Agnes M. Larson, *History of the White Pine Industry in Minnesota,* 405.

23. Beardsley, *Harry L. Russell,* 123, 126.

24. Robert Gough makes a convincing case that, despite the particular hardships of cutover farming, some settlers did succeed on their own terms. While seldom overtly prosperous, they got by on a neighborly system based on barter, growing of crops for subsistence, and occasional off-farm work. The governmental quest for rational development and social order, capped by rural zoning in the 1930s, helped push many of these economically marginal farmers off the land. See Gough, *Farming the Cutover.*

25. Richard T. Ely, "The New Economic World and the New Economics," *Journal of Land and Public Utility Economics* 5, no. 4 (November 1929): 341–53.

26. Joseph Dorfman, *The Economic Mind in American Civilization:* vol. 3, *1865–1918* (New York: Viking Press, 1949), 161–64.

27. Lafayette G. Harter Jr., *John R. Commons,* 32–35. Commons, an architect of many Progressive Era reforms, was a student of Ely's at Johns Hopkins University and later a colleague at the University of Wisconsin.

28. Leonard A. Salter Jr., *A Critical Review of Research in Land Economics*, 22–23. The institute moved to Northwestern University in 1925. The same year, it began publishing the *Journal of Land and Public Utility Economics*.

29. Richard T. Ely, "Foundations of Agricultural Prosperity," pamphlet text of speech delivered to Farm Mortgage Bankers Association of America, Kansas City, Mo., September 16, 1920, box 4, Ely papers.

30. Ely and Morehouse, *Elements of Land Economics*, 9–10.

31. Richard T. Ely, "The Passing of Laissez-Faire as Illustrated by European Experience," outline for three talks [1912], in box 8, Ely papers.

32. Ibid., 7. Tenancy in the Wisconsin cutover was low — just 5.3 percent of farmers there were tenants in 1920, compared with 14.4 percent statewide. But the cutover farmers were more encumbered by debt. "The cutover did provide settlers with land to call their own, as the promotional literature promised," Robert Gough writes, "but their hold on that land was not secure." *Farming the Cutover*, 41.

33. Ely and Morehouse, *Elements of Land Economics*, 198. In the troubled 1920s, farm tenancy would display its useful side. Unlike landowning farmers, who found it tough to liquidate their investment, tenants could simply pack their bags and move to the city.

34. Frederick Winslow Taylor, "The Principles of Scientific Management," in *Scientific Management*, 27–28. Taylor is often maligned because of his utopian faith in time-and-motion studies and his contempt for worker autonomy. But like Herbert Hoover, he saw inefficiency as the chief problem of the economy, and he genuinely believed that a better science of industry could improve the lot of the individual worker. See also Samuel Haber, *Efficiency and Uplift: Scientific Management in the Progressive Era, 1890–1920*.

35. Ely and Morehouse, *Elements of Land Economics*, 51.

36. Richard T. Ely, "Something about the Institute for Research in Land Economics and Public Utilities," n.d. [late 1920s], in box 1 of the papers of the Institute for Research in Land Economics and Public Utilities, State Historical Society of Wisconsin, Madison (hereinafter referred to at IRLEPU papers).

37. This is not to imply that governments made no effort to manage the economic aspects of natural resource use in the nineteenth century. To the contrary, governments acted decisively on two fronts. One was the provision of free or cheap land to settlers, which encouraged quick development and helped to offset the chronic shortages of labor and capital in the United States. The other was the allocation of property rights, which encouraged self-interested stewardship of natural resources and curbed the reckless waste that accompanied the unregulated exploitation of "common pool" resources such as fish and game. The lack of a modern managerial state before 1900 should not be construed as a laissez-faire approach to resource development. See James Willard Hurst, *Law and Economic Growth;* Gary D. Libecap, *Contracting for Property Rights;* Arthur F. McEvoy, *The Fisherman's Problem;* and James A. Tober, *Who Owns the Wildlife?*

38. Ely and Morehouse, *Elements of Land Economics*, 205.

39. "Joint Report of R. T. Ely and L. C. Gray," presentation for Social Sciences Research Council, Hanover, N.H., August 17–24, 1927, box 6, IRLEPU papers.

40. Richard T. Ely, typescript of "A Plea for an American Aristocracy," n.d., 23–28, in box 8, Ely papers.

41. Ely and Morehouse, *Elements of Land Economics*, 68.

42. Beardsley, *Harry L. Russell*, 131–32.

43. Gough, "Richard T. Ely," 19.

44. Salter, *A Critical Review of Research in Land Economics*, 18–19; Richard S. Kirkendall, "L.C. Gray and the Supply of Agricultural Land," *Agricultural History* 37, no. 4 (October 1963): 206–16.

45. Harry C. McDean, "Professionalism in the Rural Social Sciences, 1896–1919," *Agricultural History* 58, no. 3 (July 1984): 373–92.

46. Harry C. McDean, "'Reform' Social Darwinists and Measuring Levels of Living on American Farms, 1920–1926," *Journal of Economic History* 43, no. 1 (March 1983): 79–85. Galpin, too, had come under Ely's influence, having served on the faculty at Madison before moving to the USDA.

47. Gough, *Farming the Cutover*, 113.

3. P. S. Lovejoy and the Campaign for Order

The epigraph is from P. S. Lovejoy, "Cloverland — Watch Its Smoke!" *Country Gentleman*, March 27, 1920, 10–11, 48, 50, in U.S. Forest Service press clipping file "Cut-Over Lands," Forest History Society, Durham, N.C.

1. David Lowenthal, "The Pioneer Landscape: An American Dream," *Great Plains Quarterly* 2, no. 1 (winter 1982): 5–19.

2. Richard Hofstadter, *The Age of Reform*, 30.

3. Lowenthal, "The Pioneer Landscape," 12.

4. P. S. Lovejoy, "In the Name of Development," *American Forestry* 29 (July 1923): 387–93, 447.

5. Ibid., 390.

6. P. S. Lovejoy, "The Grazing Boom in Cloverland: Where Sure Feed and Water Are Proclaimed," *Country Gentleman* 85, no. 11 (March 13, 1920): 6–7; 64.

7. P. S. Lovejoy, "Settling the East: Michigan Tackles Job, Building from the Ground Up," *Country Gentleman* 87, no. 41 (November 18, 1922): 7; 26; P. S. Lovejoy, "Settling the East: Ballyhooing a Failure, Our States Grope for New Methods," *Country Gentleman* 87, no. 35 (October 7, 1922): 8, 32–33.

8. Richard H. Stroud, ed., *National Leaders of American Conservation*, 250; Norman J. Schmaltz, "P.S. Lovejoy: Michigan's Cantankerous Conservationist," *Journal of Forest History* 19, no. 2 (April 1975): 72–81. Lovejoy's talent for communicating high-minded ideals to ordinary citizens may have been genetically ingrained. His grandfather, Owen Lovejoy, was a firebrand abolitionist congressman from Illinois. His great-uncle, the Illinois editor Elijah Parish Lovejoy, was murdered by an angry mob for his staunch antislavery views in 1837, becoming one of the best-known martyrs to the abolitionist cause. Schmaltz, "P.S. Lovejoy," 74.

9. James Playsted Wood, *The Curtis Magazines*, 57–61, 86, 112; P. S. Lovejoy to Austin Cary, October 20, 1921, in box 4 of the P. S. Lovejoy papers, Michigan Historical Collections, Bentley Historical Library, University of Michigan, Ann Arbor (hereinafter referred to as Lovejoy papers). The *Country Gentleman* of the early 1920s was manifestly a farm magazine, not a journal for fashionably rural suburbanites. It carried ads for tractors, chick brooders, cream separators and home electrification equipment.

10. James Willard Hurst, *Law and Economic Growth*.

11. *Minnesota: Its Resources and Progress; Its Beauty, Healthfulness and Fertility; and Its Attractions and Advantages as a Home for Immigrants, with a Descriptive Map* (Saint Paul: Press Printing Company, 1870), 68, in subject file "Minnesota," Forest History Society, Durham, N.C.

12. Gavin Wright, *Old South, New South: Revolutions in the Southern Economy since the Civil War* (New York: Basic Books, 1986), 23–24. Wright defines "town building" as a northern phenomenon, saying it was largely absent in the antebellum South because local elites there held most of their capital in the form of human beings — that is, slaves. Even after the Civil War, he says, many Southerners were more interested in perpetuating a low-wage economy than they were in raising local land values through development.

13. Isaiah Bowman, *The Pioneer Fringe,* 25.

14. Leonard A. Salter Jr., *A Critical Review of Research in Land Economics,* 5–38, 83–129; J. C. McDowell and W. B. Walker, "Farming on the Cut-Over Lands of Michigan, Wisconsin, and Minnesota," USDA Bulletin 425, October 24, 1916; John Swenehart, "Clear More Land," University of Wisconsin Agricultural Experiment Station Bulletin 320, December 1920.

15. *Rhinelander (Wis.) News,* May 31, 1912.

16. P. S. Lovejoy, "For the Land's Sake: The Planting of New Forests Should Begin Without Delay," *Country Gentleman* 84, no. 30 (July 26, 1919): 8–9, 41. Earlier installments of the series ran on July 12 and 19.

17. Lovejoy, "In the Name of Development," 392.

18. Ibid., 387; P. S. Lovejoy to John E. Pickett, March 18, 1920, box 4, Lovejoy papers; P. S. Lovejoy, "Cloverland: A Part-Time Empire," *Country Gentleman* 85, no. 9 (February 28, 1920): 3–4, 42, 44.

19. P. S. Lovejoy, "It Does Not Pay to Work Land Which It Does Not Pay to Work: But Our Cut-Over Lands Can Duplicate a Crop They Once Produced," *Country Gentleman* 84, no. 29 (July 19, 1919): 10–11, 32.

20. Ibid.

21. P. S. Lovejoy, "The Promised Land: In the Northern Cut-Over Country Lies America's Great Ungrasped Opportunity," *Country Gentleman* 85, no. 51 (December 18, 1920): 3–4, 32.

22. Roderick Nash, *Wilderness and the American Mind,* 182–99.

23. P. S. Lovejoy to John E. Pickett, August 8, 1922, box 5, Lovejoy papers; P. S. Lovejoy, "Agricultural Development," unpublished manuscript, circa 1922, box 5, Lovejoy papers.

24. P. S. Lovejoy, "The Promised Land: The Wisconsin Idea in the Cut-Overs," *Country Gentleman* 86, no. 1 (January 1, 1921): 4–5, 30, 32.

25. Arlan Helgeson, *Farms in the Cutover,* 97–117.

26. P. S. Lovejoy, "State Land Policy in Michigan," paper prepared for delivery to Society of American Foresters meeting, Boston, 1922, box 4, Lovejoy papers (emphasis in original).

27. Lovejoy, "Settling the East: Ballyhooing a Failure," 32; Robert J. Gough, "Richard T. Ely and the Development of the Wisconsin Cutover," *Wisconsin Magazine of History* 75, no. 1 (autumn 1991): 3–38; Lovejoy, "Agricultural Development."

28. C. L. Harrington to P. S. Lovejoy, March 9, 1922, box 4, Lovejoy papers; A. D. Campbell to P. S. Lovejoy, September 18, 1922, box 5, Lovejoy papers; P. S. Lovejoy to A. D. Campbell, September 14, 1922, box 5, Lovejoy papers.

29. P. S. Lovejoy to John E. Pickett, March 18, 1920, box 4, Lovejoy papers; P. S. Lovejoy to Austin Cary, November 1, 1922, box 4, Lovejoy papers.

30. P. S. Lovejoy, "Settling the East: Its Empire of Idle Acres Is Increasing," *Country Gentleman* 87, no. 34 (September 30, 1922): 1–2, 16 (series continued through November 18, 1922); P. S. Lovejoy, "Farm and Forest Development in the Cutovers," paper prepared for Tri-State Development Congress, Menominee, Mich., January 18, 1923, in box 5, Lovejoy papers; *Menominee (Mich.) Herald-Leader,* January 18, 1923, clipping in box 5, Lovejoy papers.

31. P. S. Lovejoy to Austin Cary, November 1, 1922, box 4, Lovejoy papers.

32. "Parrish *[sic]* S. Lovejoy, 1884–1942," *Journal of Forestry* 40, no. 4 (April 1942): 337; Norman J. Schmaltz, "Michigan's Land Economic Survey," *Agricultural History* 52, no. 2 (April 1978): 229–46.

33. "Forest Propaganda and Forestry," *Journal of Forestry* 24, no. 4 (April 1926): 331.

34. P. S. Lovejoy to John E. Pickett, June 19, 1922, box 5, Lovejoy papers.

35. Lovejoy had no direct experience in World War I, but he did serve on a panel whose existence was due largely to war-inspired anxieties about potential shortages of timber. This was the Society of American Foresters' special committee chaired by Gifford Pinchot, whose 1919 warning of a "timber famine" is discussed in chapter 1. Schmaltz, "P.S. Lovejoy," 75–76.

36. Gough, "Richard T. Ely," 26.

37. B. H. Hibbard, John Swenehart, W. A. Hartman, and B. W. Allin, "Tax Delinquency in Northern Wisconsin," University of Wisconsin Agricultural Experiment Station Bulletin 399, June 1928; Hurst, *Law and Economic Growth,* 85. Ironically, just a few years before, John Swenehart had been the UW's leading proponent on the use of explosives for clearing cutover lands.

38. For a nuanced treatment of the tax question, see Robert Gough, *Farming the Cutover,* 145–49.

39. P. S. Lovejoy to Austin Cary, November 1, 1922, box 4, Lovejoy papers.

40. Erling D. Solberg, *New Laws for New Forests;* Hurst, *Law and Economic Growth,* 602.

41. Harold Titus, "Conservation Loses a Leader," *Michigan Conservation* 11, no. 3 (February–March 1942).

42. Norman J. Schmaltz, "Academia Gets Involved in Michigan Forest Conservation," *Michigan Academician* 12, no. 1 (summer 1979): 25–46.

43. Aldo Leopold, "Obituary: P. S. Lovejoy," *Journal of Wildlife Management* 7, no. 1 (January 1943): 125–28.

4. Cowboys and Bureaucrats

The epigraph is from Carlton Gordon Murray, "The Philosophy of the Forest," *Journal of Forestry* 26, no. 1 (January 1928): 105–9.

1. D. P. Duncan and F. H. Kaufert, "Education for the Profession," in *American Forestry,* ed. Henry Clepper and Arthur B. Meyer, 24–35.

2. "Timber cruising" is the art and science of estimating the amount and quality of timber in a forest tract by walking through it.

3. Donald C. Swain, *Federal Conservation Policy, 1921–1933,* 9–29.

4. Harold K. Steen, *The U.S. Forest Service: A History,* 196–97. Stuart's death was never officially ruled a suicide. But "[a]ccident or not, the burden of office killed Robert Stuart," Steen concludes.

5. Murray, "The Philosophy of the Forest."

6. See, for example, Will C. Barnes, "B'ar Stories," *Saturday Evening Post* 197, no. 23 (December 6, 1924): 14, 153, 157–58. Barnes's familiarity with frontier danger was more than fanciful. He had won the Medal of Honor for his service in the Indian wars.

7. Ernest Thompson Seton, *Wild Animals I Have Known.* In 1898 the author was known as Ernest Seton Thompson, but he later reversed the last two names. His whimsical depictions of animals with human qualities disturbed some naturalists and enraged Theodore Roosevelt, who branded him a "nature faker."

8. Letter to Harold Titus from editor of *Adventure* magazine (signature illegible), March 13, 1914, in box 1 of the Harold Titus papers, Michigan Historical Collections, Bentley Historical Library, University of Michigan, Ann Arbor.

9. Irving Harlow Hart, "Fiction Fashions from 1925 to 1926," *Publishers' Weekly* 111, no. 6 (February 5, 1927): 473–77. Curwood, as will be seen in chapter 5, was essentially an antimodernist. His books ran decidedly against the tone of greater realism in forest tales after 1920.

10. Roderick Nash, *The Nervous Generation,*140.

11. Paul H. Hosmer, "Forest Fires in Real Life and Reel Life," *American Forests and Forest Life* 32, no. 390 (June 1926): 323–25, 358.

12. For the essence of this argument I am indebted to Susan J. Douglas, *Inventing American Broadcasting, 1899–1922.* Douglas writes about the media celebration of the "boy-heroes" of radio, the amateur operators who constituted a cult of technical wizardry before commercial broadcasting began in 1920. Amateurs, Douglas says, were portrayed in newspaper accounts as individual geniuses on the frontiers of technical innovation. But of course, the forces of organization — in this case, the research laboratory and the modern corporation — were waiting in the wings. Radio, just like forestry circa 1920, represented "a new realm in which science and romance commingled." *Inventing American Broadcasting,* 191.

13. Steen, *The U.S. Forest Service,* 36.

14. Ibid., 62.

15. Ibid., 78–80.

16. John Milton Cooper Jr., "Gifford Pinchot Creates a Forest Service," in *Leadership and Innovation,* ed. Jameson W. Doig and Erwin C. Hargrove, 63–95.

17. Steen, *The U.S. Forest Service,* 82–83.

18. Jack Welch, "The New Forest Assistant," in *The Forest Ranger and Other Verse,* ed. John D. Guthrie, 101–2. The "cob" of the verse was a corncob pipe. Guthrie's book, which was dedicated to Gifford Pinchot, was a collection of verse from early Forest Service days. Contributors included Aldo Leopold and P. S. Lovejoy.

19. In at least one respect, the glorification of the ranger in popular culture worked all too well: in 1927, the Forest Service complained that it was being besieged by job-seekers who had been trained by ranger correspondence schools. The schools, according to forestry officials, had advertised the ranger's work as "a play-time job, or a convenient means for an outing in the woods." U.S. Department of Agriculture, Office of Information, "Forest Service Warns against Rosy Pictures Advertised

by Ranger Correspondence Schools," press release, March 21, 1927, in U.S. Forest Service press clipping file "Forest Rangers," Forest History Society, Durham, N.C.

20. E. R. Jackson, "The Forest Ranger," *American Forestry* 17, no. 8 (August 1911): 445–55.

21. Alfred Pittman, "Adventures of a Forest Ranger," *American Magazine* 100, no. 2 (August 1925): 56–59, 94, 96, 98.

22. Ovid Butler, ed., *Rangers of the Shield.*

23. Wallace Hutchinson, "Singed, but Safe: A 'Ranger Bill' Story," *American Forests and Forest Life* 36 (July 1930): 432–33. (Biographical information for Hutchinson appears on p. 463 of the same issue.)

24. U.S. Forest Service, *Service Bulletin* 13, no. 13 (April 1, 1929), in subject file "Forest Rangers," Forest History Society, Durham, N.C.

25. A. C. McCain, "Patrolling the Forest Hinterland," *American Forests and Forest Life* 36 (July 1930): 419–20.

26. "Life of a Forest Ranger," *Popular Mechanics* 44, no. 4 (October 1925): 611–15.

27. Huber C. Hilton, "Woman's Place Is in the — Mountains!", *Outing* 80, no. 1 (April 1922): 295.

28. Pittman, "Adventures of a Forest Ranger," 59, 96. Kreutzer's life and work were eventually celebrated in a full-length biography. See Len Shoemaker, *Saga of a Forest Ranger.*

29. E. C. Pulaski, "Surrounded by Fire," in *Rangers of the Shield*, 77–81. The telling of fire tales, according to one historian, also helped rangers to forge an "informal folk culture" and occupational identity within their ranks. See Timothy Cochrane, "Trial by Fire: Early Forest Service Rangers' Fire Stories," *Forest and Conservation History* 35, no. 1 (January 1991): 16–23.

30. "Life of a Forest Ranger," 614.

31. Pittman, "Adventures of a Forest Ranger," 98.

32. Jackson, "The Forest Ranger," 446.

33. Richard Slotkin, *Gunfighter Nation,* 22–26.

34. Samuel P. Hays, *Conservation and the Gospel of Efficiency,* 123.

35. A cautionary note from T. J. Jackson Lears is in order: "I want to revise Gramsci by drawing on Freud: to show that a change in cultural hegemony stems not only from deliberate persuasion by members of a dominant class but also from half-conscious hopes and aspirations which seem to have little to do with the public realm of class relations." T. J. Jackson Lears, *No Place of Grace,* xv.

36. A. J. Hanna, "A Bibliography of the Writings of Irving Bacheller," *Rollins College Bulletin* 35, no. 1 (September 1939). By Hanna's accounting, *Eben Holden* sold at least 400,000 copies. The admiring critic was Clarence Hurd Gaines, a professor of English at Saint Lawrence University, which was Bacheller's alma mater.

37. Irving Bacheller, *Silas Strong, Emperor of the Woods,* 56–57.

38. Ibid., 204.

39. Here and elsewhere, *Silas Strong* goes beyond Turnerian anxieties into the realm of outright antimodernism. The antimodernists focused, not on the point where "civilization" and "savagery" met, but on the zone of "savagery" itself, which was thought to be a repository of purifying experience including direct, continuous, mystical contact with nature. For another example of the fictional forest as dwelling place of spirits, see Elia W. Peattie, *The Beleaguered Forest.*

40. Bacheller, *Silas Strong*, 337.

41. Theodore Roosevelt Jr., *Stewart Edward White, Novelist of the American Frontier* (New York: Doubleday, Doran, n.d.). This promotional booklet was issued about 1940, when White was still active. See also Edna Rosemary Butte, "Stewart Edward White."

42. Stewart Edward White, *The Blazed Trail*. Part of the novel was serialized in *McClure's* magazine to expose the corruption of the timber business. Butte, *Stewart Edward White*. 136–37.

43. White, *The Blazed Trail*, 52.

44. Ibid., 113.

45. Ibid., 301.

46. Ibid., 3.

47. Ibid., 4.

48. Ibid., 35.

49. Butte, "Stewart Edward White," 152.

50. Theodore Roosevelt to Stewart Edward White, October 8, 1904, in *The Letters of Theodore Roosevelt*: vol. 4, *The Square Deal, 1903–1905*, ed. Elting E. Morison (Cambridge, Mass.: Harvard University Press, 1951), 977–78.

51. *Theodore Roosevelt: An Autobiography*, 31.

52. Hendryx was a friend and fishing partner of the writer Harold Titus, who is discussed in chapter 7. Like Titus, he lived in the Grand Traverse Bay region of Michigan and ran an orchard.

53. James B. Hendryx, *Connie Morgan in the Lumber Camps*. Heinie Metzger, Hendryx's fictional German timber baron, clearly was intended to mirror Frederick Weyerhaeuser.

54. This distinction is evident not only in the novel's text, but also in its illustrations.

55. Hendryx, *Connie Morgan in the Lumber Camps*, 245.

56. Ibid., 164.

57. James B. Hendryx, *Connie Morgan with the Forest Rangers*.

58. Ibid., 161.

59. Ibid., 89.

60. Ibid., 169. *Connie Morgan with the Forest Rangers* is rather baldly imitative of Harold Titus's *Timber* (1922), which will be discussed at length in chapter 7.

5. James Oliver Curwood and the Limits of Antimodernism

The second epigraph is from a letter sent by P. S. Lovejoy to Harold Titus, February 22, 1922, in box 1 of the Harold Titus papers, Michigan Historical Collections, Bentley Historical Library, University of Michigan, Ann Arbor (hereinafter referred to as Titus papers).

1. James Oliver Curwood, *God's Country*, 7.

2. Judith A. Eldridge, *James Oliver Curwood*, 2.

3. Ibid., 149.

4. Curwood, *God's Country*, 12.

5. To date, there is no satisfactory biography of Curwood. Eldridge's *James Oliver Curwood* is awkwardly written and much too admiring. Curwood's autobiography,

The Glory of Living, was first published in England about 1928. It was the basis for an American edition revised and updated by Dorothea A. Bryant, *Son of the Forests*. The autobiographies are vintage Curwood, painting his life as a romantic adventure. A worshipful tome by Hobart D. Swiggett, *James Oliver Curwood: Disciple of the Wilds*, amplifies and embellishes Curwood's own legends of himself.

6. For my understanding of antimodernism, I am relying on T. J. Jackson Lears. "The antimodern impulse stemmed from the revulsion against the process of rationalization first described by Max Weber — the systematic organization of economic life for maximum productivity... the drive for efficient control of nature under the banner of improving human welfare; the reduction of the world to a disenchanted object to be manipulated by rational technique." Antimodernism supplied a critique of all bureaucratized, rationalized systems, whether capitalist or socialist. T. J. Jackson Lears, *No Place of Grace*, 7.

7. See Stephen Fox, *The American Conservation Movement*. Fox makes sporadic, admiring mention of Curwood, particularly his pantheistic creed.

8. Swiggett, *Disciple of the Wilds*, 185.

9. Obituary in the *New York Times*, August 14, 1927.

10. Swiggett, *Disciple of the Wilds*, 102–3; Curwood and Bryant, *Son of the Forests*, 123–26.

11. James Oliver Curwood, *The Glory of Living*, 128.

12. Eldridge, *James Oliver Curwood*, 50; Curwood, *The Glory of Living*, 221.

13. Eldridge, *James Oliver Curwood*, 53, 55.

14. Swiggett, *Disciple of the Wilds*, 137.

15. See, for example, the correspondence regarding the transfer of two city lots in Saskatoon, Saskatchewan: R. R. Hartney to Curwood, July 29, 1909, in box 1 of the James Oliver Curwood papers, Michigan Historical Collections, Bentley Historical Library, University of Michigan, Ann Arbor (hereinafter referred to as Curwood papers).

16. James Oliver Curwood, *Steele of the Royal Mounted*, 9–10.

17. Ibid., 186.

18. Eldridge, *James Oliver Curwood*, 89.

19. Curwood, *The Glory of Living*, 28–31.

20. Curwood, *God's Country*, 99–101.

21. James Oliver Curwood, "The God of Her People," *Pearson's Magazine* 26, no. 6 (December 1911): 775–780+, incomplete clipping in box 3, Curwood papers.

22. Curwood, *God's Country*, 13.

23. James Oliver Curwood, *The Bear: A Novel*. Director Jean-Jacques Annaud adapted Curwood's tale for the critically acclaimed 1988 film *The Bear*.

24. Ibid., 166, 168. In the novel, Curwood suggests that bears have the capacity for love, friendship, paternal care, and loyalty.

25. Ibid., 26.

26. Acting as his own producer, Curwood took enormous risks — artistically and financially — to make screen versions of his books such as *Back to God's Country* and *Nomads of the North*. After the early 1920s he left the moviemaking to others.

27. Curwood to Alex. W. Bissland, December 12, 1921, box 1, Curwood papers; clipping from *Flint Daily Journal*, December 10, 1921, in scrapbook accompanying Curwood papers (hereinafter referred to as Curwood scrapbook).

28. *Detroit Free Press*, July 16, 1921.

29. Curwood to T. F. Marston, July 15, 1921, box 1, Curwood papers.

30. Curwood to J. W. Tracy, January 5, 1922, box 1, Curwood papers.

31. Curwood to W. A. Brewer, January 2, 1922, box 1, Curwood papers. A little math and some basic silviculture belie Curwood's alarm. Planted at five-foot intervals, a million Christmas trees would occupy slightly less than a square mile (640 acres). Assuming that a tree needed nine years to reach cutting size, a plot just three miles square would generate a million trees a year in perpetuity. It would provide employment, tax revenue, and a modicum of game habitat and scenery in the process. Properly managed, a Christmas-tree farm was a microcosm of the cutover revival as many planners envisioned it.

32. P. S. Lovejoy to Curwood, December 14, 1921, box 1, Curwood papers.

33. Curwood to P. S. Lovejoy, December 15, 1921, box 1, Curwood papers.

34. Curwood to P. S. Lovejoy, n.d. [August 1922], box 1, Curwood papers.

35. P. S. Lovejoy to Curwood, February 4, 1922, box 1, Curwood papers.

36. P. S. Lovejoy to Harold Titus, February 22, 1922, box 1, Titus papers. In speech and in written correspondence, Curwood consistently referred to reforestation as "reforestration."

37. P. S. Lovejoy to Curwood, August 22, 1922, box 1, Curwood papers.

38. Fred W. Green to Curwood, March 23, 1922; Curwood to Fred W. Green, March 25, 1922, both in box 1, Curwood papers.

39. Alva M. Cummins to Curwood, October 1, 1922, box 1, Curwood papers.

40. Curwood to Filibert Roth, October 31, 1922, box 1, Curwood papers. Roth was one of Curwood's more steadfast supporters in the scientific community. He served on the Michigan Conservation Commission but quit in 1922, disgusted with the political infighting that came with the job.

41. Curwood's charges against Groesbeck and Baird appear to have been about half correct. Groesbeck, according to his biographer, was a brusque and forceful leader who nonetheless possessed certain conciliatory skills. Conservation Director Baird was not a trained scientist. But Eugene Thor Petersen, a historian of the period, says that Baird—though chosen for his "political affiliations"—was "far superior" to his predecessors in conservation work. Curwood's mistake lay not in criticizing Groesbeck and Baird, but in assuming that nothing could be done until they were gone. Frank B. Woodford, *Alex J. Groesbeck,* 10, 126–27, 129, 137; Norman John Schmaltz, "Cutover Land Crusade," 116, 188, 439–67; Eugene Thor Petersen, "The History of Wild Life Conservation in Michigan," 259, 264.

42. Curwood's involvement with the "Ike Waltons" would be noteworthy for his fiery defense of Will Dilg, the passionate founder of the national group. Dilg would be accused of malfeasance and deposed by the league in 1926. Typically, Curwood rushed into the fracas and became Dilg's fervent supporter. Curwood quit the league in protest, to the accompaniment of much publicity. Eventually, as the evidence against Dilg mounted, Curwood stepped back from the battle. Dilg died of cancer in March 1927.

43. Ivan A. Conger, *Curwood Castle,* undated brochure issued by Owosso Historical Commission. Curwood Castle is now open to the public as a tourist attraction.

44. Ray Long, "James Oliver Curwood and His Far North," *Bookman* 52, no. 6 (February 1921): 492–95.

45. "On Superior's Shore," *New York Times Book Review,* July 27, 1924, 18.

46. Written before Curwood found his voice as a wilderness author, *Captain Plum* was a historical novel set among the Mormons on Beaver Island in Lake Michigan.

47. James Oliver Curwood, *A Gentleman of Courage.*

48. Ibid., 101.

49. Ibid., 214.

50. Ibid., 256.

51. Ibid., 265.

52. Ibid., 21–22.

53. Ibid., 100.

54. Ibid., 20. Of all organized religions, Curwood professed the most admiration for Catholicism. There were two reasons for this. First, Catholicism was the creed of the Jesuit missionaries and many French-Canadian pioneers, so the faith was tied up with notions of adventure and romance. Second, like many antimodernists, Curwood embraced the Catholic Church's sense of mystery and pageantry as an alternative to Protestant sects' increasing rationality and worldliness after Darwin. Lears, *No Place of Grace,* 184–86.

55. Curwood, *A Gentleman of Courage,* 146.

56. Curwood to Ray Long, n.d. [c. 1919], box 1, Curwood papers. Long had suggested that Curwood try writing farm stories.

57. James Oliver Curwood, *The Ancient Highway.*

58. Ibid., 124–25.

59. Richard Slotkin, *Gunfighter Nation,* 216–17.

60. Fred Green to Curwood, January 4, 1927, box 2, Curwood papers.

61. *Grand Rapids (Mich.) Herald,* January 6, 1927, Curwood scrapbook.

62. Michigan Public Act 17 of 1921, quoted in "The Department of Conservation: The Formative Years," undated manuscript history on file at Michigan Department of Natural Resources, Lansing, 250.

63. Michigan Department of Conservation, *Proceedings of Conservation Commission, January 19, 1927.* Here, Curwood was following the lead of his heroes in the animal-protection movement, such as William Temple Hornaday, who believed that the simplest way to protect animals was to stop their harvest by sportsmen. Other thinkers, such as Aldo Leopold, were starting to realize that fish and game management was far more complicated, involving such matters as habitat, interdependence of species, and naturally occurring cycles of plenty and scarcity among game animals. But the Hornaday view was more popular during the 1920s.

64. Curwood to Harold Titus, January 25, 1927, box 2, Curwood papers.

65. Curwood to Fred Green, January 22, 1927, box 2, Curwood papers.

66. *Detroit Times,* February 2, 1927, Curwood scrapbook.

67. *Detroit News,* June 9, 1927, Curwood scrapbook.

68. *Detroit Free Press,* April 27, 1927.

69. William B. Mershon to Leigh J. Young, April 27, 1927, in box 1 of the Leigh J. Young papers, Michigan Historical Collections, Bentley Historical Library, University of Michigan, Ann Arbor.

70. Michigan Department of Conservation, *Proceedings of Conservation Commission, May 4–5, 1927; Michigan State Digest,* May 14, 1927, Curwood scrapbook.

71. *Detroit Free Press,* May 6, 1927; Curwood to Harold Titus, May 27, 1927, box 2, Curwood papers.

72. *Detroit News,* June 9, 1927, Curwood scrapbook.

73. *Manistee (Mich.) News-Advocate,* July 7, 1927, Curwood scrapbook.

74. *Detroit Free Press,* July 7, 1927, Curwood scrapbook.

75. *Manistee News-Advocate,* July 7, 1927, Curwood scrapbook.

76. *Gratiot County (Mich.) Herald,* July 14, 1927, Curwood scrapbook.

77. Lee Smits to Curwood, July 8, 1927; Curwood to Lee Smits, July 11, 1927, box 2, Curwood papers.

78. *Detroit Free Press,* August 4, 1927.

79. Eldridge, *James Oliver Curwood,* 215, 224–27.

80. *Detroit Free Press,* September 8, 1927. The action on fish spearing appears to have been a direct rebuff to Curwood. In the months before his death, the author had produced a series of articles decrying the spear as a "barbed destroyer."

81. Michigan Department of Conservation, *Proceedings of Conservation Commission, September 7, 1927.*

82. James Oliver Curwood, *Green Timber,* completed by Dorothea A. Bryant.

83. Ibid., 125. Writers in this era often played up the cutover's reputation as a hideout for gangsters, adding to the popular impression that the region was "backward and dangerous," as well as lawless. Robert Gough, *Farming the Cutover,* 150–51.

84. Curwood, *Green Timber,* 77.

85. Ibid., 121.

86. Ibid., 282.

87. Curwood's ethos was essentially biocentric, meaning that he saw nature, not man, as the center of existence. Most 1920s conservationists, such as P. S. Lovejoy, conceived of the natural world in terms of what it could do for human beings. Man, Curwood wrote, "is the human peacock, puffed up, inflated, flushed in the conviction *that everything in the universe was made for him.*" Curwood, *God's Country,* 9–10 (emphasis in original).

6. The Production of Leisure

The epigraph is from Herbert Hoover, "In Praise of Izaak Walton," *Atlantic Monthly* 139 (June 1927): 813–19.

1. William Voigt Jr., "The Izaak Walton League of America," *Journal of Forestry* 44, no. 6 (June 1946): 424–25.

2. Zane Grey, "Vanishing America," clipping from *Izaak Walton League Monthly* 1, no. 2 (September 1922), no page given, in box 1 of the James Oliver Curwood papers, Michigan Historical Collections, Bentley Historical Library, University of Michigan, Ann Arbor. The league's magazine was renamed *Outdoor America* in October 1923.

3. Henry S. Graves, "A Crisis in National Recreation," *American Forestry* 26, no. 319 (July 1920): 391–400.

4. Ibid., 391.

5. L. F. Kneipp, "Forest Recreation Comes of Age," *American Forests and Forest Life* 36 (July 1930): 415–18.

6. For an example of the anxieties of Progressive Era recreation experts, see John Collier, "Leisure Time, the Last Problem of Conservation," *Playground* 6, no. 3 (June 1912): 93–106.

7. "National Conference on Outdoor Recreation," U.S. Senate Document no. 158 (Washington: U.S. Government Printing Office, 1928), 5. The National Conference, convened by President Calvin Coolidge in 1924, was a wide-ranging coalition of planners, foresters, playground experts, municipal reformers, and others who met and issued recommendations on outdoor recreation, to the accompaniment of heavy publicity. For a comprehensive report on the conference's first meeting, see *Playground* 18, no. 4 (July 1924).

8. Hoover, "In Praise of Izaak Walton." Hoover epitomized all the energy and ironies of the New Era conservation movement. While invoking the carefree fishing days of his boyhood in Iowa, he simultaneously viewed angling as a problem best attacked through administrative and scientific means. See Carl E. Krog, "'Organizing the Production of Leisure': Herbert Hoover and the Conservation Movement in the 1920s," *Wisconsin Magazine of History* 67, no. 3 (spring 1984): 199–218; and Kendrick A. Clements, "Herbert Hoover and Conservation, 1921–1933," *American Historical Review* 89, no. 1 (February 1984): 67–88.

9. J. F. Steiner, "Recreation and Leisure Time Activities," in U.S. Office of the President, *Recent Social Trends in the United States,* 2: 912–57.

10. U.S. Department of Agriculture, Forest Service, "Vacation in the National Forests" (Washington: U.S. Government Printing Office, 1921), in box 72 of the Society of American Foresters papers, Forest History Society, Durham, N.C. (hereinafter referred to as SAF papers).

11. Alexander Johnston, "America: Touring Ground of the World," *Country Life* 37, no. 3 (January 1920): 25–34.

12. "Living in the Car," *Outing* 82, no. 1 (April 1923): 18, 46–47.

13. Advertisement for "The Burch Auto Bed and Tent," *Outing* 78, no. 2 (May 1921): 88.

14. Lawrence S. Clark, "Six Weeks in a Ford," *Outing* 80, no. 4 (July 1922): 162–64.

15. Frank A. Waugh, "The Public Road — Our Great National Park," *House Beautiful,* June 1920, 508–9, 528, in U.S. Forest Service press clipping file "Roads," Forest History Society, Durham, N.C. (hereinafter referred to as USFS clipping file, by subject).

16. J. C. Long, "The Motor Car as the Missing Link between Country and Town," *Country Life* 43, no. 4 (February 1923): 112.

17. George S. Wehrwein, "Some Problems of Recreational Land," *Journal of Land and Public Utility Economics* 3, no. 2 (May 1927): 163–72.

18. U.S. Department of Agriculture, Office of the Secretary, "Over 8,000,000 Automobilists Visit Uncle Sam's Forests," press release, March 11, 1924, in USFS clipping file "Recreation."

19. Herbert Evison, "The Problem of the Gypsy Automobilist," *American Forests and Forest Life* 31, no. 377 (May 1925): 273–74.

20. S. Herbert Hare, "The Relation of the Landscape Architect to the State Park Movement," *Landscape Architecture* 16, no. 4 (July 1926): 222–29.

21. C. P. Halligan, "Tourist Camps," Michigan Agricultural Experiment Station Special Bulletin 139 (East Lansing: Michigan Agricultural College, 1925), 12, 17.

22. Allen S. Peck, "The Camper's Trust," *Road,* August 1924, 15, 17, in USFS clipping file "Recreational Forestry."

23. Erle Kauffman, "Camping as She Is Did," *American Forests and Forest Life* 33, no. 403 (July 1927): 401–3, 448.

24. Harry Irving Shumway, "Club Connections for Motor Campers," *American Forests and Forest Life* 30, no. 370 (October 1924): 581–84. Ads for chain-link and wooden fences appeared frequently in forestry and garden periodicals during the 1920s, often with pointed references to their role in discouraging campers and picnickers. A 1926 ad for Anchor Fences, for example, pictured a clan unloading a flivver and spreading picnic debris all over the lawn of a country estate. "An Anchor Fence is a permanent and diplomatic solution to this annoying problem," the copy stated. "It is a sure-to-be-obeyed, but with-a-smile, way of commanding— 'Keep Off!' " Ad in *American Forests and Forest Life* 32, no. 386 (February 1926): 109.

25. "Outdoor Recreation Resources of the United States Federal Lands," report by joint committee of the American Forestry Association and the National Parks Association to the National Conference on Outdoor Recreation (1926), 141–44, typescript in box 71, SAF papers.

26. Shumway, "Club Connections," 584.

27. Linda Flint McClelland, *Presenting Nature: The Historic Landscape Design of the National Park Service, 1916 to 1942* (Washington: U.S. Department of the Interior, National Park Service, 1993), 161–66.

28. W. B. Mershon, "When There Were Turkeys in Michigan," *Outing* 80, no. 2 (May 1922): 58–62. The wild turkey has been restored through scientific management and is once again a popular game bird in Michigan and elsewhere.

29. Ibid., 62.

30. Ovid Butler, "The Conservation of Wildlife," *American Forests* 41, no. 9 (September 1935): 482–87.

31. Nineteenth-century wildlife protection efforts are outlined in James A. Tober, *Who Owns the Wildlife?*

32. Aldo Leopold, "A History of Ideas in Game Management," *Outdoor America* 9, no. 11 (June 1931): 22–24, 38–39, 47.

33. Ibid., 47. Game management lagged about three decades behind forestry in its development as an academic discipline and as a profession. Leopold, for example, lacked a textbook for the teaching of game courses at the University of Wisconsin, so he wrote one himself: the classic *Game Management* (1933).

34. Aldo Leopold, "The National Forests: The Last Free Hunting Grounds of the Nation," *Journal of Forestry* 17, no. 2 (February 1919): 150–53.

35. Any mere summary of Leopold's thought is inadequate. The reader seeking a definitive treatment is directed to Curt Meine, *Aldo Leopold: His Life and Work.*

36. Aldo Leopold, "Game Management in the National Forests," *American Forests and Forest Life* 36 (July 1930): 412–13.

37. Meine, *Aldo Leopold,* 294.

38. C. E. Rachford, "Game Administration on the National Forests," speech delivered to Fifteenth National Game Conference, New York City, December 3–4, 1928, text in box 1 of the Harold K. Steen U.S. Forest Service research files, Forest History Society, Durham, N.C.

39. P. S. Lovejoy to Aldo Leopold, July 27, 1928, in series 9/25/10–3, box 6, of the Aldo Leopold papers, University of Wisconsin–Madison Archives (hereinafter referred to as Leopold papers).

40. Besides manipulating animal habitat, feed, and mortality patterns, managers also could alter the animals themselves, through selective breeding and release in the wild. Paul G. Redington, chief of the U.S. Bureau of Biological Survey, suggested

as much in 1929, questioning whether "nature has done all that can be done in the development or improvement of our native wild life species." Through hybridization, animals might be made "more attractive for the chase or the table," he said. U.S. Department of Agriculture, Office of Information, "Redington Sees Profits from Wild Life in Forests," press release, February 5, 1929, in USFS clipping file "Game."

41. Ben East, "Where Shall We Hunt?" *Outdoor Life/Outdoor Recreation,* March 1931, 24–25+ (clipping incomplete), in series 9/25/10–3, box 6, Leopold papers.

42. "The Farmer and the Sportsman: Jack Van Coevering Explains the Famous Williamston Plan," unidentified, undated clipping [c. 1931] in series 9/25/10–3, box 6, Leopold papers.

43. Earl C. Doyle, "Williamston National Project: Progress Report No. 1," January 21, 1931, in series 9/25/10–3, box 6, Leopold papers.

44. Earl C. Doyle, "Getting to the Bottom of Conservation through the Williamston Wild Life Project," *Magazine of Michigan,* February 1931, 5, 24+ (clipping incomplete), in series 9/25/10–3, box 6, Leopold papers. Institutions participating in the project included Michigan State College, the University of Michigan, the Michigan Conservation Department, the Michigan Academy of Science, and the U.S. Bureau of Biological Survey.

45. Doyle, "Williamston National Project: Progress Report No. 1," 4.

46. "Realization of League Aims," Earl C. Doyle memo to Aldo Leopold et al., n.d. [1931], in series 9/25/10–3, box 6, Leopold papers.

47. Voigt, "The Izaak Walton League of America."

48. Grey, "Vanishing America" (emphasis in original).

49. Stephen Fox, *The American Conservation Movement,* 159–63. Dilg was deposed as president by the league after a power struggle in 1926, allegedly for expense-account abuses. He died of cancer the following year. Wm. A. Bruette, "The Izaak Walton League Convention," *Forest and Stream* 96, no. 6 (June 1926): 327–30.

50. "Radios Line Up; W.G.E.S. Is Walton Station," *Walton League Bulletin* 1, no. 9 (January 1, 1928): 4, in box 2 of the Izaak Walton League of Wisconsin papers, State Historical Society of Wisconsin, Madison (hereinafter referred to as IWLW papers). Waltonians in Milwaukee took to the air as early as 1925, and a popular Walton program over WCCO radio in Minneapolis featured tales of the legendary lumberjack Paul Bunyan.

51. Will H. Dilg, "The Drainage Crime of a Century," reprint from *Izaak Walton League Monthly* 1, no. 11 (July 1923) in box 1, IWLW papers.

52. Henry Baldwin Ward, "Fish — and More Fish," *Outdoor America* 7, no. 3 (October 1928): 16–17, 39, 54. Fish culture actually was one of the older conservation sciences, with federal support dating from 1871. But until about 1930, the rearing and stocking of fish tended to be done somewhat blindly, with little attention paid to habitat and other ecological complexities. Ovid Butler, "The Conservation of Fish Life," *American Forests* 41, no. 9 (September 1935): 488–93.

53. "Save the Starving Birds — Feed Them!" bulletin from Izaak Walton League of America headquarters to all chapters, December 28, 1926, in box 2, IWLW papers.

54. Haskell Noyes to "Fellow Waltonians," May 14, 1925, box 1, IWLW papers.

55. "A Battle the Izaak Walton League Is Winning," cartoon by John T. McCutcheon, n.d. [c. 1927], box 2, IWLW papers.

56. Dilg, "The Drainage Crime of a Century."

57. Clipping from *Chicago Tribune,* November 6, 1923, in box 1, IWLW papers.

58. Will H. Dilg to "Brother Waltonian," March 11, 1925, box 1, IWLW papers.

59. The chief result of this corporate-funded work was Leopold's *Report on a Game Survey of the North Central States* (Madison, Wis., 1931). The American Game Protective Association also was funded largely by donations from arms and ammunition makers. Perhaps reflecting that influence, it actively supported the propagation of game, not just its protection.

60. *The Out-of-Doors Appeal in Advertising: A Contribution to the Art of Building Advertisements by "Field and Stream"* (New York: *Field and Stream* magazine, n.d. [c. 1923]).

61. Advertisement in *American Magazine* 98, no. 3 (September 1924): 81.

62. Advertisement in *Garden and Home Builder* 43, no. 6 (August 1926): back cover.

63. Advertisement in *Outdoor America* 7, no. 5 (December 1928): inside back cover.

64. David Backes, "Wilderness Visions: Arthur Carhart's 1922 Proposal for the Quetico-Superior Wilderness," *Forest and Conservation History* 35, no. 3 (July 1991): 128–37.

65. This meeting, which took place in Denver, is recounted in Meine, *Aldo Leopold*, 177–78.

66. Arthur H. Carhart, "Memorandum for Mr. Leopold, District 3," December 10, 1919, copy in series 9/25/10–11, box 2, Leopold papers.

67. Backes, "Wilderness Visions," 132.

68. Aldo Leopold, "The Wilderness and Its Place in Forest Recreational Policy," *Journal of Forestry* 19, no. 7 (November 1921): 718–21.

69. Aldo Leopold, "Wilderness as a Form of Land Use," *Journal of Land and Public Utility Economics* 1, no. 4 (October 1925): 398–404.

70. Ibid., 398.

71. Ibid., 400.

72. Aldo Leopold, "The Last Stand of the Wilderness," *American Forests and Forest Life* 31, no. 382 (October 1925): 599–604. For a thoughtful reply to Leopold's proposal, see Howard R. Flint, "Wasted Wilderness," *American Forests and Forest Life* 32, no. 391 (July 1926): 407–10, with Leopold comment at 410–11.

73. Leopold, "Wilderness as a Form of Land Use," 401.

74. Manly Thompson, "A Call from the Wilds," U.S. Forest Service, *Service Bulletin*, 12, no. 20 (May 14, 1928): 2–3, with typescript of Leopold reply, "Mr. Thompson's Wilderness," June 6, 1928, both in series 9/25/10–4, box 8, Leopold papers.

75. Donald N. Baldwin, *The Quiet Revolution*, 8–9. The first officially designated large-scale wilderness in the national forests, chosen on authority of the district forester, was the Gila Wilderness of Arizona and New Mexico, in 1924.

76. James M. Glover, "Romance, Recreation, and Wilderness: Influences on the Life and Work of Bob Marshall," *Environmental History Review* 14, no. 4 (winter 1990): 22–39.

7. Harold Titus

The epigraph is from Ben East, "Titus Prophet of Conservation for Michigan's Wilds," unidentified clipping [c. 1933] in box 5 of the Harold Titus papers, Michigan Historical Collections, Bentley Historical Library, University of Michigan, Ann Arbor (hereinafter referred to as Titus papers).

1. Harold Titus, "The Picture That Walked," *American Forestry* 28 (December 1922): 715–19.

2. Harold Titus to E. G. Rich, March 2, 1922, box 1, Titus papers.

3. Manuscript fragment of autobiography, box 3, Titus papers. Titus wrote eleven novels and hundreds of stories and articles for magazines such as *Collier's* and *Red Book*. He served on the Michigan Conservation Commission for nearly twenty years. He died in 1967.

4. "Titus Plans Romance of the Michigan Fur Trade," *Midland Michigan*, October 31, 1935, in box 5, Titus papers; file of *Detroit News* clippings, 1908–10, box 5, Titus papers. Titus received an honorary master's degree at Ann Arbor in 1931.

5. Diary of cowboy experience, n.d. [c. 1912], box 5, Titus papers.

6. Herbert S. Case, ed., *The Official Who's Who in Michigan* (Munising, Mich.: Who's Who in Michigan, 1936), 395. Titus's increasing prominence as a writer is documented by miscellaneous business correspondence in box 1, Titus papers.

7. R. H. Davis to Harold Titus, March 12, 1914, box 1, Titus papers.

8. Harold Titus to "Mr. Hale," December 19, 1917, box 1, Titus papers. The proposed novel apparently was never written.

9. Harold Titus, "It Can't Be Done," manuscript story [c. 1919] in box 4, Titus papers.

10. Manuscript fragment of autobiography, box 3, Titus papers.

11. "The Gossip Shop," *Bookman* 55, no. 4 (June 1922): 448.

12. Advertisement in *Publishers' Weekly* 101, no. 12 (March 25, 1922): 894.

13. Harold Titus, *Timber*.

14. Harold Titus to E. G. Rich, July 30, 1921, box 1, Titus papers.

15. Clarence Andrews, *Michigan in Literature*, 88–89; East, "Titus Prophet of Conservation for Michigan's Wilds."

16. Titus, *Timber*, 122.

17. Ibid., 30.

18. One of Helen Foraker's few local allies is a journalist — Humphrey "Hump" Bryant, state senator and editor of the *Blueberry Banner*. Bryant is working for timberland property-tax reform, which in 1922 was a very real issue in the Great Lakes states. Reduction or deferral of property taxes on forest lands removed an incentive for the landowner to cut the timber before it was truly "ripe" in an economic sense.

19. *Timber*, 59–62.

20. Ibid., 319–20.

21. Ibid., 338. In another tribute to the mythical value of the forest, Titus's novel also includes one of the first Paul Bunyan stories to be published in a book. (Titus spelled the name "Bunion.") It is noteworthy that the Paul Bunyan tales, which had been spun by loggers for decades, did not appear in print until the Great Lakes forest was gone — when writers were striving to articulate a rationale for the forest's rebirth. Andrews, *Michigan in Literature*, 89.

22. Titus's interest in conservation doubtless cost him much writing income after 1920 and especially after 1927, when he served on the Michigan Conservation Commission. Unlike his fellow author-conservationist James Oliver Curwood, Titus could not afford the diversion. His papers show that he suffered considerable financial trouble and was forced to borrow money at times.

23. Harold Titus to E. G. Rich, July 8, 1921, box 1, Titus papers.

24. Michael Williams, *Americans and Their Forests,* 443–46. The most tangible result of this fervor on the federal level was the Clarke-McNary Act of 1924, which set a standard for federal-state cooperation and funding for fire protection.

25. P. S. Lovejoy to Harold Titus, January 28, 1922, box 1, Titus papers.

26. Gifford Pinchot to P. S. Lovejoy, March 23, 1922, copy in box 1, Titus papers.

27. Huber C. Hilton to Harold Titus, July 29, 1922, box 1, Titus papers.

28. Ezra Levin to Harold Titus, May 1, 1922, box 1, Titus papers.

29. J. A. Doelle to Harold Titus, April 28, 1922, box 1, Titus papers. In praising *Timber,* Doelle displayed a certain generosity of spirit. Just two years before, he had been secretary of the Upper Peninsula Development Bureau, whose agricultural schemes Lovejoy had lampooned in his "Cloverland" articles for the *Country Gentleman.* Doelle quickly came around on the forestry question, and he and Lovejoy came to share a mutual respect.

30. P. S. Lovejoy to Harold Titus, September 18, 1922, box 1, Titus papers.

31. Quoted in Andrews, *Michigan in Literature,* 89.

32. P. S. Lovejoy to G. M. Householder, March 12, 1923, copy in box 1, Titus papers. See *Hearts Aflame,* motion picture copyright description no. LP18571, at Motion Picture, Broadcasting and Recorded Sound Division, Library of Congress, Washington, D.C.

33. P. S. Lovejoy memo for H. J. Andrews, January 18, 1942, copy in "Speeches" file, box 5, Titus papers. This memo, written just two days before Lovejoy's death, was one of several in which he summed up the high points of conservation history as he saw it.

34. Titus's language here is reminiscent of that of the land economists, who asserted that a "proper" use existed for each acre and that science could reveal that use. Such a conception overlooked the cultural and economic determinants of "proper" land use, as the more candid land economists came to admit. Likewise, in fish and game management, the "proper" solution to any problem would vary over time, given variances in power relationships and the perceived desirability of particular goals.

35. Author's foreword to "Leaves from the Old Warden's Notebook," with promotional brochure, n.d. [1934] in box 5, Titus papers.

36. "The Old Warden on Suckers," undated manuscript [c. 1933], box 3, Titus papers.

37. "The Old Warden on Chicken," undated manuscript [c. 1933], box 3, Titus papers. "Prairie chicken" is a generic term for certain types of grouse. In other articles, the "Old Warden" ruminated about trout habitat, diseases among deer, ring-necked pheasants (a nonnative species imported from Asia, which came to be a staple of Midwestern hunting), and other topics.

8. Spiritual Means to Economic Ends

The epigraphs are from E. T. Allen, "50,000 Firebrands," reprint from *American Forests and Forest Life,* n.d. [c. 1926], in box 57 of the American Forestry Association papers, Forest History Society, Durham, N.C. (hereinafter referred to as AFA papers); and John D. Guthrie, "Fire in the Sanctuary," reprint from *Outdoor Amer-*

ica, August 1928, in U.S. Forest Service press clipping file "Forest Fires," Forest History Society, Durham, N.C. (hereinafter referred to as USFS clipping file, by subject).

1. Bernhard Fernow, quoted in Stephen J. Pyne, *Fire in America,* 165.

2. P. S. Lovejoy, "Cloverland—Watch Its Smoke!" *Country Gentleman,* March 27, 1920, 10–11, 48, 50, in USFS clipping file "Cut-Over Lands."

3. J. A. Mitchell and H. R. Sayre, *Forest Fires in Michigan* (Lansing: Michigan Department of Conservation, 1931), 8. Great Lakes fires flared up again during the drought years of the early 1930s.

4. Michael Williams, *Americans and Their Forests,* 481–83. Overall forest-fire damage peaked in 1931, with 53 million acres burned.

5. Ovid Butler, "The War against Forest Fires," *American Forests* 41, no. 9 (September 1935): 464–69.

6. Stewart H. Holbrook, *Burning an Empire,* 194–95.

7. Lovejoy, "Cloverland—Watch Its Smoke!" 50.

8. Holbrook, *Burning an Empire,* 61–74. The disaster at Peshtigo received less press coverage than it might have, for two reasons. First, Peshtigo was remotely located. It was several weeks, for example, before *Harper's Weekly* managed to get a correspondent and a sketch artist into the area. Second, Peshtigo burned on the very same night, at nearly the same hour, as did Chicago. The great Chicago Fire took only one-fifth as many lives, but got far more attention in newspapers and magazines. The fire at Chicago also left more than 100,000 people homeless, an element of drama that was lacking at Peshtigo.

9. Transcript of interview with Scott Leavitt, July 6, 1960, p. 28, in Oral History Collection, Forest History Society, Durham, N.C.

10. H. C. Putnam, "Forest Fires," *American Journal of Forestry* 1, no. 1 (October 1882): 27–30, in the George Wirt collection on the early history (1873–97) of the American Forestry Association, at Forest History Society, Durham, N.C. (emphasis in original).

11. Lawrence H. Larsen, *Wall of Flames,* 32, 155, 159.

12. E. G. Cheyney, "The Holocaust in Minnesota: A Greater Hinckley," *American Forestry* 24, no. 299 (November 1918): 643–47, clipping in box 563 of the Gifford Pinchot papers, Library of Congress, Washington, D.C.

13. Holbrook, *Burning an Empire,* 39–45. The Great Lakes states did experience large fires after Cloquet, but the loss of life was far less. Specifically, 1925 was a bad year, with 1.4 million acres burning in Minnesota, Wisconsin, and Michigan. Pyne, *Fire in America,* 200.

14. U.S. Department of Agriculture, Forest Service, *The Use Book,* 95.

15. Pyne, *Fire in America,* 189. It is noteworthy that many old-time forest rangers— the men with cowboy roots who were hired at the end of the nineteenth century— did not share their younger colleagues' total aversion to fire. In the eyes of these veterans, fire could be both useful and destructive. But their beliefs were overridden by a new generation of college-educated, politically aware foresters, for whom total fire exclusion became a professional creed. See Timothy Cochrane, "Trial by Fire: Early Forest Service Rangers' Fire Stories," *Forest and Conservation History* 35, no. 1 (January 1991): 16–23.

16. Pyne, *Fire in America,* 167.

17. C. L. Harrington memo on forest-fire situation (headed "Dear Sir"), January 19, 1921, in box 1 of the C. L. Harrington papers, State Historical Society of Wisconsin, Madison (hereinafter referred to as Harrington papers).

18. Wisconsin Conservation Commission (signed by C. L. Harrington) to "Town Chairmen in Northern Counties," March 27, 1922, box 1, Harrington papers.

19. Anita Bowden to Mrs. D. J. Evans, November 25, 1925, letter on file at State Historical Society of Wisconsin, Madison.

20. Pyne, *Fire in America*, 169.

21. Erle Kauffman, "Camera! Foresters Turn to Grease Paint and Powder to Further Educational Work in South," *American Forests and Forest Life* 35, no. 10 (October 1929): 619–22, 648.

22. Ovid Butler, "Office Memorandum: Trip on Account Southern Forestry Educational Project," November 6, 1929, in Ovid Butler manuscript file, Forest History Society, Durham, N.C.

23. U.S. Department of Agriculture, Forest Service, *The Forest Situation in the United States: A Special Report to the Timber Conservation Board* (January 30, 1932), 20, in box 27, Society of American Foresters papers, Forest History Society, Durham, N.C.

24. "Southern Forestry Educational Project of the American Forestry Association," memo probably written by Ovid Butler [1931], in box 57, AFA papers.

25. "Material for Conservation Commission Conference on Industrial Forestry," February 12, 1946, typescript in box 5, Harrington papers; Neil H. LeMay, "History of Forest Fire Protection in Wisconsin," *Proceedings of Tenth Annual Meeting of Forest History Association of Wisconsin, Inc.* (1985), 29–35.

26. Russell Watson, "Memorandum for Mr. Stoll [Albert Stoll Jr., secretary of Michigan Conservation Commission] — Forest Fires," October 14, 1922, in box 3 of the James Oliver Curwood papers, Michigan Historical Collections, Bentley Historical Library, University of Michigan, Ann Arbor (hereinafter referred to as Curwood papers). Watson was a forestry professor at the University of Michigan.

27. Ibid.

28. Chas. R. Meek, "Forest Fire Education and Publicity Methods," *Journal of Forestry* 21, no. 3 (March 1923): 242–47.

29. Donald Hough, "The Canoe Fire Department," *Outing* 79, no. 3 (December 1921): 107–9.

30. U.S. Department of Agriculture, Office of the Secretary, "Army of 22,000,000 to Combat Forest Fires," press release, November 13, 1922, in USFS clipping file "Forest Fires."

31. Shirley W. Allen, *The Forest Fire Helpers: A Masque* (Washington, D.C.: American Forestry Association, n.d. [c. 1928]), in box 57, AFA papers.

32. Meek, "Forest Fire Education and Publicity Methods," 246.

33. Robert Shaler, *The Boy Scouts as Forest Fire Fighters*, 40, 43 (emphasis in original).

34. Guthrie, "Fire in the Sanctuary." Guthrie was editor of *The Forest Ranger and Other Verse*, cited in chapter 4.

35. Aldo Leopold, "Wild Followers of the Forest," *American Forestry* 29, no. 357 (September 1923): 515–19, 568.

36. E. T. Allen, "50,000 Firebrands."

37. Paul H. Hosmer, "Forest Fires in Real Life and Reel Life," *American Forests and Forest Life* 32, no. 390 (June 1926): 323–25, 358.

38. Joseph Montrose to James Oliver Curwood, May 19, 1920, box 1, Curwood papers.

39. Pyne, *Fire in America*, 196.

40. Wallace I. Hutchinson, "Public Relations: What Have We Bought and Where Are We Headed?" *Journal of Forestry* 29, no. 4 (April 1931): 474–83. Hutchinson was the author of the "Ranger Bill" tales, described in chapter 4.

41. Erle Kauffman, "Progressive Publicity in Forestry," *American Forests and Forest Life* 34, no. 410 (February 1928): 93–95.

42. Ward Shepard, "The Necessity for Realism in Forestry Propaganda," *Journal of Forestry* 25, no. 1 (January 1927): 11–26 (emphasis in original).

43. Professional mass communication talent would not be enlisted for the forest-fire crusade until the 1940s, when the Foote, Cone & Belding ad agency would create Smokey Bear.

9. Rural Zoning and the Synthetic Frontier

The epigraph is from W. A. Rowlands, "County Zoning for Agriculture, Forestry, and Recreation in Wisconsin," *Journal of Land and Public Utility Economics* 9, no. 3 (August 1933): 272–82, quoting John W. Reynolds.

1. "Theodore Roosevelt Publicizes Conservation, 1908," in *Major Problems in American Environmental History*, ed. Carolyn Merchant, 350–52.

2. "The New Patriotism," June 18, 1908, speech drafts in box 18 of the George S. Wehrwein papers, State Historical Society of Wisconsin, Madison (hereinafter referred to as Wehrwein papers); undated clipping from *Oshkosh Democrat* (July 1908), in box 18, Wehrwein papers.

3. Samuel Lubell and Walter Everett, "Wisconsin Revives the Wilderness," *Reader's Digest* 35, no. 209 (September 1939): 59–62. A longer version of this article had appeared a month earlier in *Current History*, a magazine that examined social issues from a New Deal perspective.

4. F. G. Wilson, "Zoning for Forestry and Recreation: Wisconsin's Pioneer Role," *Wisconsin Magazine of History* 41, no. 2 (winter 1957–58): 102–6.

5. Lubell and Everett, "Wisconsin Revives the Wilderness."

6. George S. Wehrwein, "The Farm Tenant of the South," *Public Affairs*, April 1924, 11, 14, in box 18, Wehrwein papers.

7. George S. Wehrwein, "What Are the Facts about Arable Land?" *National Real Estate Journal* 24, no. 2 (January 15, 1923): 13–14, in box 18, Wehrwein papers.

8. P. S. Lovejoy, "Theory and Practice in Land Classification," *Journal of Land and Public Utility Economics* 1, no. 2 (April 1925): 160–75. "Safe" in this context apparently meant safe from fire.

9. Wade DeVries, "The Michigan Land Economic Survey," *Journal of Farm Economics* 10, no. 4 (October 1928): 516–24.

10. Richard H. D. Boerker, "Continuous Forest Production and Rural Independence," undated manuscript [c. 1939], in box 1 of the Richard H. D. Boerker papers, Forest History Society, Durham, N.C.

11. O. B. Jesness and R. I. Nowell, "Zoning of Minnesota Lands," University of Minnesota Agricultural Extension Division Special Bulletin 167 (August 1934): 3.

12. DeVries, "The Michigan Land Economic Survey," 516–17.

13. George S. Wehrwein, "For What Good Is Land Anyway?" script for radio talk on WHA, November 6, 1930, box 18, Wehrwein papers.

14. Peter Hall, ed., *Von Thünen's Isolated State.*

15. Isaiah Bowman, *The Pioneer Fringe,* vi, 49.

16. George S. Wehrwein, "Rural Problems in Our Marginal Areas," attachment with letter to Ernest Burnham, October 22, 1930, box 3, Wehrwein papers.

17. Ibid.; George S. Wehrwein, "A Social and Economic Program for the Sub-Marginal Areas of the Lake States," *Journal of Forestry* 29, no. 6 (October 1931): 915–24.

18. Wehrwein, "A Social and Economic Program," 917; N. Cornutt to University of Wisconsin Agricultural Experiment Station, June 9, 1932; George S. Wehrwein to N. Cornutt, June 15, 1932, both in box 5, Wehrwein papers.

19. Walter A. Rowlands, "A Collection of Some of the Most Interesting Stories, Incidents and Statements Given by Local Officials and Residents of Northern Wisconsin on the Value and Significance of the Rural Zoning Movement in Wisconsin," mimeograph typescript (Madison, 1945), in pamphlet collection, State Historical Society of Wisconsin, Madison.

20. Erling D. Solberg, *New Laws for New Forests,* 265–69. Some counties in southern Wisconsin enacted rural zoning ordinances as well. But the southern ordinances merely laid out patterns of land use (agricultural, industrial, residential), whereas cutover zoning was aimed specifically at controlling the distribution of the human population.

21. Solberg, *New Laws for New Forests,* 332.

22. Robert Gough, *Farming the Cutover,* 168–69. Gough says that, while relief was a fact of life in the Depression-era cutover, proportionally fewer farmers got relief checks than did cutover citizens in other lines of work. *Farming the Cutover,* 131.

23. *Rhinelander (Wis.) Daily News,* January 18, May 6, 1933.

24. Rowlands, "County Zoning for Agriculture, Forestry, and Recreation," 281. From the time city zoning became popular in the 1920s, courts generally were unsympathetic to the argument that zoning constituted an unconstitutional "taking" of private property. See, for example, *Village of Euclid et al. v. Ambler Realty Company,* 272 U.S. 365 (1926).

25. *Rhinelander Daily News,* May 17, 1933.

26. George S. Wehrwein, "Resettlement and Rehabilitation," script for radio talk on WHA, October 28, 1935, box 18, Wehrwein papers. Zoning ordinances permitted fines or jail terms for people who moved into restricted regions after the laws were passed. In most instances, community pressure apparently sufficed to persuade illegal settlers to give up their farms. In 1940, a municipal judge in Bayfield County sentenced a couple, Roy and Rose Johannes, to twenty days in jail for violating the zoning ordinance there. Rowlands, "A Collection of Some of the Most Interesting Stories."

27. "The New Patriotism," speech drafts in box 18, Wehrwein papers; Paul A. Eke to George S. Wehrwein, August 12, 1929, and George S. Wehrwein to Paul A. Eke, August 20, 1929, both in box 2, Wehrwein papers.

28. George S. Wehrwein, "Why Do We Need to Zone Land?" script for radio talk on WHA, September 4, 1933, box 18, Wehrwein papers.

29. George S. Wehrwein and J. A. Baker, "The Cost of Isolated Settlement in Northern Wisconsin," *Rural Sociology* 2, no. 3 (September 1937): 253–65. Much of this work was carried out in preparation for a federally sponsored resettlement project that never was completed. Resettlement — the voluntary, government-sponsored moving of isolated settlers into more populated areas with better soils — turned out to be a fairly minor component of the cutover revival. About four hundred Wisconsin families had been resettled by 1939. Lubell and Everett, "Wisconsin Revives the Wilderness," 60.

30. W. C. Nichols to George S. Wehrwein, January 29, 1934; Louis W. Plost to George S. Wehrwein, January 30, 1934, both in box 7, Wehrwein papers.

31. Wehrwein and Baker, "The Cost of Isolated Settlement," 258.

32. Lubell and Everett, "Wisconsin Revives the Wilderness."

33. Robert J. Gough, "Richard T. Ely and the Development of the Wisconsin Cutover," *Wisconsin Magazine of History* 75, no. 1 (autumn 1991): 3–38.

Epilogue

1. Rexford Guy Tugwell, "Down to Earth," *Current History* 44, no. 4 (July 1936): 33–38.

2. "Forests and Land Use," Wisconsin Department of Agriculture Bulletin 229 (April 1942): 15.

3. Tugwell, "Down to Earth," 35.

4. David B. Danbom, *The Resisted Revolution*, 139.

5. Ibid., 128.

6. Samuel P. Hays, *Beauty, Health, and Permanence*, 528.

7. Aldo Leopold, "Conservation Esthetic," in *A Sand County Almanac and Sketches Here and There*, 165–77.

Bibliography

Manuscript Collections

Forest History Society, Durham, North Carolina

American Forestry Association. Papers.
Boerker, Richard H. D. Papers.
Butler, Ovid M. Manuscript file.
Butler, Ovid M. Publications file.
Leavitt, Scott. Oral history transcript.
Society of American Foresters. Papers.
Steen, Harold K. U.S. Forest Service research files.
Subject files: Biltmore Estate and Forest School, Forest Fires, Forest Rangers, Michigan, Minnesota, Wisconsin.
United States Forest Service press clipping file.
Wirt, George. Collection on early history (1873–97) of the American Forestry Association.

Library of Congress, Washington, D.C.

Motion picture copyright descriptions.
Pinchot, Gifford. Papers.

Michigan Department of Natural Resources, Lansing

"The Department of Conservation: The Formative Years." Manuscript history on file at DNR headquarters.

Michigan Historical Collections, Bentley Historical Library, University of Michigan, Ann Arbor

Curwood, James Oliver. Papers and clipping scrapbook.
Hazzard, Albert Sidney. Papers.
Lovejoy, Parish Storrs. Papers.
Titus, Harold. Papers.
Young, Leigh J. Papers.

State Historical Society of Wisconsin, Madison, Archives Division

Bordner, John S. Papers.
Bowden, Anita. Letter, 1925.
Ely, Richard T. Papers.
Harrington, C. L. Papers.
Institute for Research in Land Economics and Public Utilities. Papers.
Izaak Walton League of Wisconsin. Papers.
Luther, Ernest L. Papers.
Wehrwein, George S. Papers.

University of Wisconsin–Madison Archives

Leopold, Aldo. Papers.

Nonfiction Books as Primary Sources

Bailey, L. H. *The Country-Life Movement in the United States.* New York: Macmillan, 1911.
———. *The Garden Lover.* New York: Macmillan, 1928.
———. *The Harvest of the Year to the Tiller of the Soil.* New York: Macmillan, 1927.
———. *The Holy Earth.* New York: Scribner, 1915.
Bowman, Isaiah. *The Pioneer Fringe.* New York: American Geographical Society, 1931.
Butler, Ovid, ed. *Rangers of the Shield.* Washington, D.C.: American Forestry Association, 1934.
Curwood, James Oliver. *The Glory of Living: The Autobiography of an Adventurous Boy Who Grew into a Writer and a Lover of Life.* Mattituck, N.Y.: Aeonian Press, 1983.
———. *God's Country: The Trail to Happiness.* New York: Cosmopolitan Book Corporation, 1921.
Curwood, James Oliver, and Dorothea A. Bryant. *Son of the Forests.* Garden City, N.Y.: Doubleday, Doran, 1930.
Ely, Richard T., and Edward W. Morehouse. *Elements of Land Economics.* New York: Macmillan, 1924.
Galpin, Charles Josiah. *Rural Life.* New York: Century, 1922.
Hawthorn, Horace Boies. *The Sociology of Rural Life.* New York: Century, 1926.
Hibbard, Benjamin Horace. *A History of the Public Land Policies.* New York: Macmillan, 1924.
Holmes, Roy Hinman. *The Farm in a Democracy.* Ann Arbor, Mich.: Edwards, 1922.
Leopold, Aldo. *Game Management.* New York: Scribner, 1933.
———. *Report on a Game Survey of the North Central States.* Madison, Wis.: Sporting Arms and Ammunition Manufacturers' Institute, 1931.
Lippmann, Walter. *Drift and Mastery.* New York: Mitchell Kennerley, 1914.
MacKaye, Benton. *The New Exploration: A Philosophy of Regional Planning.* Urbana: University of Illinois Press, 1962.
Mershon, Wm. B. *Recollections of My Fifty Years Hunting and Fishing.* Boston: Stratford, 1923.
The Out-of-Doors Appeal in Advertising: A Contribution to the Art of Building Advertisements by "Field and Stream." New York: *Field and Stream* magazine, n.d. [c. 1923].

Pack, Charles Lathrop, and Tom Gill. *Forests and Mankind.* New York: Macmillan, 1929.

Pinchot, Gifford. *The Fight for Conservation.* New York: Doubleday, Page, 1910.

Roosevelt, Theodore. *Theodore Roosevelt: An Autobiography.* New York: Scribner, 1921.

Taylor, Frederick Winslow. *Scientific Management.* New York: Harper, 1947.

Turner, Frederick Jackson. *The Frontier in American History.* New York: Holt, 1920.

Waugh, Frank A. *The Natural Style in Landscape Gardening.* Boston: Badger/Gorham Press, 1917.

Nonfiction Books as Secondary Sources

Ackerman, Bruce A., ed. *Economic Foundations of Property Law.* Boston: Little, Brown, 1975.

Albanese, Catherine L. *Nature Religion in America: From the Algonkian Indians to the New Age.* Chicago: University of Chicago Press, 1990.

Alchon, Guy. *The Invisible Hand of Planning: Capitalism, Social Science, and the State in the 1920s.* Princeton, N.J.: Princeton University Press, 1985.

Anderson, H. Allen. *The Chief: Ernest Thompson Seton and the Changing West.* College Station: Texas A&M University Press, 1986.

Andrews, Clarence. *Michigan in Literature.* Detroit, Mich.: Wayne State University Press, 1992.

Backes, David James. "The Communication-Mediated Roles of Perceptual, Political, and Environmental Boundaries on Management of the Quetico-Superior Wilderness of Ontario and Minnesota, 1920–1965." Ph.D. dissertation, University of Wisconsin–Madison, 1988.

Baldwin, Donald N. *The Quiet Revolution: Grass Roots of Today's Wilderness Preservation Movement.* Boulder, Colo.: Pruett, 1972.

Beardsley, Edward H. *Harry L. Russell and Agricultural Science in Wisconsin.* Madison: University of Wisconsin Press, 1969.

Bogue, Allan G., and Robert Taylor, eds. *The University of Wisconsin: One Hundred and Twenty-Five Years.* Madison: University of Wisconsin Press, 1975.

Bowers, William L. *The Country Life Movement in America, 1900–1920.* Port Washington, N.Y.: Kennikat Press, 1974.

Butte, Edna Rosemary. "Stewart Edward White: His Life and Literary Career." Ph.D. dissertation, University of Southern California, 1960.

Carhart, Arthur H. *The National Forests.* New York: Knopf, 1959.

Carstensen, Vernon. *Farms or Forests: Evolution of a State Land Policy for Northern Wisconsin, 1850–1932.* Madison: University of Wisconsin College of Agriculture, 1958.

Clepper, Henry, and Arthur B. Meyer, eds. *American Forestry: Six Decades of Growth.* Washington, D.C.: Society of American Foresters, 1960.

Cox, Thomas R., Robert S. Maxwell, Phillip Drennon Thomas, and Joseph J. Malone. *This Well-Wooded Land: Americans and Their Forests from Colonial Times to the Present.* Lincoln: University of Nebraska Press, 1985.

Cronon, William. *Nature's Metropolis: Chicago and the Great West.* New York: Norton, 1992.

———, ed. *Uncommon Ground: Toward Reinventing Nature.* New York: Norton, 1995.

Danbom, David B. *The Resisted Revolution: Urban America and the Industrialization of Agriculture, 1900–1930.* Ames: Iowa State University Press, 1979.

Doig, Jameson W., and Erwin C. Hargrove, eds. *Leadership and Innovation: A Biographical Perspective on Entrepreneurs in Government.* Baltimore, Md.: Johns Hopkins University Press, 1987.

Douglas, Susan J. *Inventing American Broadcasting, 1899–1922.* Baltimore, Md.: Johns Hopkins University Press, 1989.

Dunbar, Willis Frederick. *Michigan: A History of the Wolverine State,* 2nd ed. Grand Rapids, Mich.: Eerdmans, 1970.

Eldridge, Judith A. *James Oliver Curwood: God's Country and the Man.* Bowling Green, Ohio: Bowling Green State University Popular Press, 1993.

Eyle, Alexandra. *Charles Lathrop Pack: Timberman, Forest Conservationist, and Pioneer in Forest Education.* Syracuse, N.Y.: ESF College Foundation, 1992.

Flader, Susan L. *Thinking Like a Mountain: Aldo Leopold and the Evolution of an Ecological Attitude toward Deer, Wolves, and Forests.* Columbia: University of Missouri Press, 1974.

———, ed. *The Great Lakes Forest: An Environmental and Social History.* Minneapolis: University of Minnesota Press, 1983.

Fox, Richard Wightman, and T. J. Jackson Lears, eds. *The Culture of Consumption: Critical Essays in American History, 1880–1980.* New York: Pantheon, 1983.

Fox, Stephen. *The American Conservation Movement: John Muir and His Legacy.* Madison: University of Wisconsin Press, 1981.

Friedman, Lawrence M. *A History of American Law.* 2nd ed. New York: Touchstone/Simon and Schuster, 1985.

Gallant, Christine, ed. *Coleridge's Theory of Imagination Today.* New York: AMS Press, 1989.

Gough, Robert. *Farming the Cutover: A Social History of Northern Wisconsin, 1900–1940.* Lawrence: University Press of Kansas, 1997.

Graham, Otis L., Jr. *Toward a Planned Society: From Roosevelt to Nixon.* New York: Oxford University Press, 1976.

Greeley, William B. *Forests and Men.* Garden City, N.Y.: Doubleday, 1951. Reprint, New York: Arno Press, 1972.

Grese, Robert E. *Jens Jensen: Maker of Natural Parks and Gardens.* Baltimore, Md.: Johns Hopkins University Press, 1992.

Guttenberg, Albert Z. *The Language of Planning: Essays on the Origins and Ends of American Planning Thought.* Urbana: University of Illinois Press, 1993.

Haber, Samuel. *Efficiency and Uplift: Scientific Management in the Progressive Era, 1890–1920.* Chicago: University of Chicago / Midway Reprints, 1973.

Hall, Peter, ed. *Von Thünen's Isolated State: An English Edition of "Der Isolierte Staat" by Johann Heinrich von Thünen.* Oxford: Pergamon Press, 1966.

Harter, Lafayette G., Jr. *John R. Commons: His Assault on Laissez-Faire.* Corvallis: Oregon State University Press, 1962.

Hawley, Ellis W. *The Great War and the Search for a Modern Order: A History of the American People and Their Institutions, 1917–1933.* New York: St. Martin's Press, 1979.

———, ed. *Herbert Hoover as Secretary of Commerce: Studies in New Era Thought and Practice.* Iowa City: University of Iowa Press, 1981.

Hays, Samuel P. *Beauty, Health, and Permanence: Environmental Politics in the United States, 1955–1985.* Cambridge, England: Cambridge University Press, 1987.

————. *Conservation and the Gospel of Efficiency: The Progressive Conservation Movement, 1890–1920.* Cambridge, Mass.: Harvard University Press, 1959.

Helgeson, Arlan. *Farms in the Cutover: Agricultural Settlement in Northern Wisconsin.* Madison: State Historical Society of Wisconsin, 1962.

Hidy, Ralph W., Frank Ernest Hill, and Allan Nevins. *Timber and Men: The Weyerhaeuser Story.* New York: Macmillan, 1963.

Hoffman, Daniel. *Paul Bunyan: Last of the Frontier Demigods.* Lincoln: Bison Book/ University of Nebraska Press, 1983.

Hofstadter, Richard. *The Age of Reform: From Bryan to F.D.R.* New York: Vintage, 1955.

Holbrook, Stewart H. *Burning an Empire: The Story of American Forest Fires.* New York: Macmillan, 1943.

Hurst, James Willard. *Law and Economic Growth: The Legal History of the Lumber Industry in Wisconsin, 1836–1915.* Cambridge, Mass.: Belknap/Harvard University Press, 1964.

Huthmacher, J. Joseph, and Warren I. Susman, eds. *Herbert Hoover and the Crisis of American Capitalism.* Cambridge, Mass.: Schenkman, 1973.

Jackson, J. B. [John Brinckerhoff]. *Landscapes: Selected Writings of J.B. Jackson.* Ed. Ervin H. Zube. Amherst: University of Massachusetts Press, 1970.

Jackson, John Brinckerhoff. *Discovering the Vernacular Landscape.* New Haven, Conn.: Yale University Press, 1984.

Kirschner, Don S. *City and Country: Rural Responses to Urbanization in the 1920s.* Westport, Conn.: Greenwood Press, 1970.

Kloppenberg, James T. *Uncertain Victory: Social Democracy and Progressivism in European and American Thought, 1870–1920.* New York: Oxford University Press, 1986.

Kolehmainen, John I., and George W. Hill. *Haven in the Woods: The Story of the Finns in Wisconsin.* Madison: State Historical Society of Wisconsin, 1951.

Larsen, Lawrence H. *Wall of Flames: The Minnesota Forest Fire of 1894.* Fargo: North Dakota Institute for Regional Studies, 1984.

Larson, Agnes M. *History of the White Pine Industry in Minnesota.* 1949. Reprint, New York: Arno Press, 1972.

Lears, T. J. Jackson. *No Place of Grace: Antimodernism and the Transformation of American Culture, 1880–1920.* New York: Pantheon, 1981.

Levy, David W. *Herbert Croly of The New Republic: The Life and Thought of an American Progressive.* Princeton, N.J.: Princeton University Press, 1985.

Libecap, Gary D. *Contracting for Property Rights.* Cambridge, England: Cambridge University Press, 1989.

Luccarelli, Mark. *Lewis Mumford and the Ecological Region.* New York: Guilford Press, 1995.

Marchand, Roland. *Advertising the American Dream: Making Way for Modernity, 1920– 1940.* Berkeley: University of California Press, 1985.

Marx, Leo. *The Machine in the Garden: Technology and the Pastoral Ideal in America.* New York: Oxford University Press, 1964.

McCraw, Thomas K., ed. *Regulation in Perspective: Historical Essays.* Cambridge, Mass.: Harvard University Press, 1981.

McEvoy, Arthur F. *The Fisherman's Problem: Ecology and Law in the California Fisheries, 1850–1980.* Cambridge, England: Cambridge University Press, 1986.

Meine, Curt. *Aldo Leopold: His Life and Work.* Madison: University of Wisconsin Press, 1988.

Merchant, Carolyn, ed. *Major Problems in American Environmental History: Documents and Essays.* Lexington, Mass.: Heath, 1993.

Meyer, Roy W. *The Middle Western Farm Novel in the Twentieth Century.* Lincoln: University of Nebraska Press, 1965.

Miller, Donald L. *Lewis Mumford: A Life.* New York: Weidenfeld & Nicolson, 1989.

Mowry, George E. *The Era of Theodore Roosevelt and the Birth of Modern America, 1900–1912.* New York: Harper Torchbook, 1962.

Namorato, Michael V. *Rexford G. Tugwell: A Biography.* New York: Praeger, 1988.

Nash, Roderick. *The Nervous Generation: American Thought, 1917–1930.* Chicago: Rand McNally, 1970.

———. *Wilderness and the American Mind.* 3rd ed. New Haven, Conn.: Yale University Press, 1982.

Nash, Roderick Frazier. *The Rights of Nature: A History of Environmental Ethics.* Madison: University of Wisconsin Press, 1989.

Neth, Mary. *Preserving the Family Farm: Women, Community, and the Foundations of Agribusiness in the Midwest, 1900–1940.* Baltimore, Md.: Johns Hopkins University Press, 1995.

Petersen, Eugene Thor. "The History of Wild Life Conservation in Michigan." Ph.D. dissertation, University of Michigan, 1952.

Pinchot, Gifford. *Breaking New Ground.* New York: Harcourt, Brace, 1947.

Pinkett, Harold T. *Gifford Pinchot: Private and Public Forester.* Urbana: University of Illinois Press, 1970.

Pollan, Michael. *Second Nature: A Gardener's Education.* New York: Atlantic Monthly Press, 1991.

Pyne, Stephen J. *Fire in America: A Cultural History of Wildland and Rural Fire.* Princeton, N.J.: Princeton University Press, 1982.

Reiger, John F. *American Sportsmen and the Origins of Conservation.* Norman: University of Oklahoma Press, 1986.

Roper, Laura Wood. *FLO: A Biography of Frederick Law Olmsted.* Baltimore, Md.: Johns Hopkins University Press, 1973.

Runte, Alfred. *National Parks: The American Experience.* Lincoln: University of Nebraska Press, 1979.

Salter, Leonard A., Jr. *A Critical Review of Research in Land Economics.* Madison: University of Wisconsin Press, 1967.

Schenck, Carl Alwin. *The Biltmore Story: Recollections of the Beginning of Forestry in the United States.* St. Paul: American Forest History Foundation and Minnesota Historical Society, 1955.

Schmaltz, Norman John. "Cutover Land Crusade: The Michigan Forest Conservation Movement, 1899–1931." Ph.D. dissertation, University of Michigan, 1972.

Schmitt, Peter J. *Back to Nature: The Arcadian Myth in Urban America.* New York: Oxford University Press, 1969.

Schneirov, Matthew. *The Dream of a New Social Order: Popular Magazines in America, 1893–1914.* New York: Columbia University Press, 1994.

Searle, R. Newell. *Saving Quetico-Superior: A Land Set Apart.* St. Paul: Minnesota Historical Society Press, 1977.

Shoemaker, Len. *Saga of a Forest Ranger.* Boulder: University of Colorado Press, 1958.

Slotkin, Richard. *Gunfighter Nation: The Myth of the Frontier in Twentieth-Century America*. New York: Atheneum, 1992.

Smith, Henry Nash. *Virgin Land: The American West as Symbol and Myth*. 1950. Reprint, New York: Vintage Books, 1957.

Solberg, Erling D. *New Laws for New Forests: Wisconsin's Forest-Fire, Tax, Zoning, and County-Forest Laws in Operation*. Madison: University of Wisconsin Press, 1961.

Steen, Harold K. *The U.S. Forest Service: A History*. Seattle: University of Washington Press, 1976.

Stroud, Richard H., ed. *National Leaders of American Conservation*. Washington, D.C.: Smithsonian Institution Press, 1985.

Susman, Warren I. *Culture as History: The Transformation of American Society in the Twentieth Century*. New York: Pantheon, 1984.

Swain, Donald C. *Federal Conservation Policy, 1921–1933*. Berkeley: University of California Press, 1963.

Swanson, Merwin Robert. "The American Country Life Movement, 1900–1940." Ph.D. dissertation, University of Minnesota,1972.

Swiggett, Hobart D. *James Oliver Curwood: Disciple of the Wilds*. New York: Paebar, 1943.

Tober, James A. *Who Owns the Wildlife? The Political Economy of Conservation in Nineteenth-Century America*. Westport, Conn.: Greenwood Press, 1981.

Tuan, Yi-Fu. *Space and Place: The Perspective of Experience*. London: Edward Arnold, 1979.

Wadland, John Henry. *Ernest Thompson Seton: Man in Nature and the Progressive Era, 1880–1915*. New York: Arno Press, 1978.

Wiebe, Robert H. *The Search for Order, 1877–1920*. New York: Hill and Wang, 1967.

Willey, Basil. *Samuel Taylor Coleridge*. New York: Norton Library, 1973.

Williams, Michael. *Americans and Their Forests: A Historical Geography*. Cambridge, England: Cambridge University Press, 1989.

Wilson, Joan Hoff. *Herbert Hoover: Forgotten Progressive*. Boston: Little, Brown, 1975.

———, ed. *The Twenties: The Critical Issues*. Boston: Little, Brown, 1972.

Wood, James Playsted. *The Curtis Magazines*. New York: Ronald Press, 1971.

Woodford, Frank B. *Alex J. Groesbeck: Portrait of a Public Man*. Detroit, Mich.: Wayne State University Press, 1962.

Worster, Donald. *The Wealth of Nature: Environmental History and the Ecological Imagination*. New York: Oxford University Press, 1993.

Novels

Bacheller, Irving. *Silas Strong, Emperor of the Woods*. New York: Harper, 1906.

Curwood, James Oliver. *The Ancient Highway: A Novel of High Hearts and Open Roads*. New York: Cosmopolitan Book, 1925.

———. *The Bear* (originally *The Grizzly King*, 1916). New York: Newmarket Press, 1989.

———. *A Gentleman of Courage: A Novel of the Wilderness*. New York: Cosmopolitan Book, 1924.

———. *The River's End: A New Story of God's Country*. New York: Grosset & Dunlap, 1919.

—————. *Steele of the Royal Mounted* (originally *Philip Steele*, 1911). New York: Triangle Books, 1938.

—————. *The Valley of Silent Men: A Story of the Three River Country.* New York: Grosset & Dunlap, 1920.

Curwood, James Oliver, completed by Dorothea A. Bryant. *Green Timber.* New York: Doubleday, Doran, 1930.

Hendryx, James B. *Connie Morgan in the Lumber Camps.* New York: Putnam, 1919.

—————. *Connie Morgan with the Forest Rangers.* New York: Putnam, 1925.

Peattie, Elia W. *The Beleaguered Forest.* New York: Appleton, 1901.

Rolvaag, O.E. *Giants in the Earth: A Saga of the Prairie.* New York: Harper, 1927.

Shaler, Robert. *The Boy Scouts as Forest Fire Fighters.* New York: Hurst & Company, 1915.

Titus, Harold. *Flame in the Forest.* New York: Burt, 1933.

—————. *The Man from Yonder.* New York: Burt, 1928.

—————. *Spindrift.* Garden City, N.Y.: Doubleday, Page, 1925.

—————. *Timber.* Boston: Small, Maynard & Co., 1922.

White, Stewart Edward. *The Blazed Trail.* New York: McClure, Phillips, 1902.

Wister, Owen. *The Virginian: A Horseman of the Plains.* New York: Macmillan, 1902.

Poetry and Essays

Grayson, David (Ray Stannard Baker). *Adventures in Contentment.* New York: Grosset & Dunlap, 1907.

Guthrie, John D., ed. *The Forest Ranger and Other Verse.* Boston: Badger/Gorham Press, 1919.

Leopold, Aldo. *A Sand County Almanac and Sketches Here and There.* New York: Oxford University Press, 1949.

Thompson, Ernest Seton. *Wild Animals I Have Known.* New York: Scribner, 1898.

Contemporaneous Newspapers, Periodicals, and Journals

American Forestry (later *American Forests and Forest Life* and *American Forests*)
The American Magazine
The Country Gentleman
Country Life in America (later *Country Life*)
Current History
Detroit Free Press
The Detroit News
Forest and Stream
Garden and Home Builder
The Garden Magazine
House and Garden
Journal of Farm Economics
Journal of Forestry
Journal of Land and Public Utility Economics
Landscape Architecture
The Milwaukee Journal
The New York Times
Outdoor America

Outing
The Playground
Rural Sociology
The Saturday Evening Post
The World's Work

Government Documents

Michigan Agricultural Experiment Station. "Tourist Camps," by C. P. Halligan. Special Bulletin 139. East Lansing: Michigan Agricultural College, 1925.

Michigan Department of Conservation. "Forestry Conference, November 28, 1922." Lansing: Department of Conservation, 1922.

Michigan Department of Conservation. *Forest Fires in Michigan,* by J. A. Mitchell and H. R. Sayre. Lansing: Department of Conservation, 1931.

Michigan Department of Conservation. *Proceedings of Conservation Commission,* various dates 1927.

University of Minnesota Agricultural Extension Division. "Zoning of Minnesota Lands," by O. B. Jesness and R. I. Nowell. Special Bulletin 167, August 1934.

University of Wisconsin Agricultural Experiment Station. "Tax Delinquency in Northern Wisconsin," by B. H. Hibbard et al. Bulletin 399, June 1928.

University of Wisconsin Agricultural Experiment Station. "Farms Follow Stumps," by H. L. Russell. Bulletin 332, April 1921.

University of Wisconsin Agricultural Experiment Station. "Clear More Land," by John Swenehart. Bulletin 320, December 1920.

University of Wisconsin College of Agriculture Extension Service. "What Chance Has a City Man on a Wisconsin Farm?" Radio circular, November 1932.

University of Wisconsin–Madison. Institute for Environmental Studies. *Images of the Cutover: A Historical Geography of Resource Utilization in the Lake Superior Region, 1845–1930,* by Charles G. Mahaffey and Felice R. Bassuk. Lake Superior Project, Center for Geographic Analysis. RF Monograph 76–15, IES report 98, June 1978.

U.S. Commission on Country Life. *Report of the Commission on Country Life.* New York: Sturgis & Walton, 1911.

U.S. Department of Agriculture. "Farming on the Cut-Over lands of Michigan, Wisconsin, and Minnesota," by J. C. McDowell and W. B. Walker. USDA Bulletin 425, October 24, 1916.

U.S. Department of Agriculture. "Land Settlement and Colonization in the Great Lakes States," by John D. Black and L. C. Gray. USDA Bulletin 1295, March 23, 1925.

U.S. Department of Agriculture. Forest Service. *The Use Book: Regulations and Instructions for the Use of the National Forest Reserves.* Washington, D.C.: U.S. Government Printing Office, 1906.

U.S. Department of the Interior. National Park Service. *Presenting Nature: The Historic Landscape Design of the National Park Service, 1916 to 1942,* by Linda Flint McClelland. Washington: U.S. Department of the Interior, 1993.

U.S. Office of the President. *Recent Social Trends in the United States: Report of the President's Research Committee on Social Trends.* 2 vols. 1933. Reprint, Westport, Conn.: Greenwood Press, 1970.

U.S. Senate. "National Conference on Outdoor Recreation." U.S. Senate Document no. 158. Washington, D.C.: U.S. Government Printing Office, 1928.

Wisconsin Department of Agriculture. "Forests and Land Use." Bulletin 229, April 1942.

Wisconsin Department of Agriculture. "The Land Economic Inventory of Northern Wisconsin: What It Is and What It Can Be Used For," by Walter A. Duffy et al. Bulletin 97, March 1929.

Index

James Kates, a longtime journalist, is an editor at the *Milwaukee Journal Sentinel* and has also worked at the *Philadelphia Inquirer* and the former *Milwaukee Journal.* He earned a Ph.D. in mass communication, with an emphasis on media history, from the University of Wisconsin–Madison, and has published articles in *American Journalism,* the *Michigan Historical Review,* and the *Wisconsin Magazine of History.*